Anonymous

Proceedings of Central Ohio Scientific Association of Urbana,

Ohio

Anonymous

Proceedings of Central Ohio Scientific Association of Urbana, Ohio

ISBN/EAN: 9783337419745

Printed in Europe, USA, Canada, Australia, Japan

Cover: Foto ©berggeist007 / pixelio.de

More available books at **www.hansebooks.com**

EEDINGS

—OF—

RAL OHIO

URBANA, OHIO.

VOL. 1–PART 1.

PUBLISHED BY THE ASSOCIATION.

URBANA, OHIO:
SAXTON & BRAND, PRINTERS
1878

TABLE OF CONTENTS.

Report of Committee

— of —

Publication.

The Committee appointed to prepare a volume of the transactions of the Association for publication, respectfully present the following : Part 1, Vol. 1. Proceedings of the Central Ohio Scientific Association, as the result of their labors. In performing the duty assigned them, the committee has selected such papers as have been read from time to time before the Association and recommended for publication, and they have also prepared a preliminary statement of the organization, membership, constitution, donations, etc., together with an abstract from the minutes of what has been done since the foundation of the society, and of the work proposed to be undertaken in the future Of the preliminary statement the committee have decided to issue an extra edition of 300 copies for the Association.

The illustrations of the work have been made with great care and regard to accuracy, and are all from original drawings. Of these Plates 1 to 8 inclusive, were executed by Mr. D. H. Sherman. The drawings for Plates 9 to 12 inclusive, were prepared by Prof. J. E. Werren. Those illustrating the Meteorological report of Milo G. Williams, Esq., were prepared by himself. All of these gentlemen have donated their work, thus materially reducing the cost of publication.

$$\left.\begin{array}{l}\text{Thos. F. Moses,}\\ \text{Chas. G. Smith,}\end{array}\right\} \textit{Committee.}$$

ON the evening of October 20th, 1874, the following persons met at the office of Dr. R. H. BOAL, Urbana, O., for the purpose of organizing a scientific association, viz: T. N. GLOVER, of Woodstock, R. H. BOAL, J. F. MEYER, T. F. MOSES, W. F. LEAHY, P. R. BENNETT, Jr., of Urbana, and L. C. HERRICK, of Woodstock, O.

T. F. MOSES was called to the chair and W. F. LEAHY chosen Secretary *pro tem.* A committee of three was appointed to draft a constitution and by-laws. After due deliberation the committee reported, and the constitution submitted by them was unanimously adopted. The Society then proceeded to effect a permanent organization by the choice of the following officers to serve for one year: President, T. N. GLOVER; Vice President, P. R. BENNETT, Jr.; Recording and Corresponding Secretary, W. F. LEAHY; Treasurer and Curator, T. F. MOSES. At the ensuing meeting, changes were made in the above list of officers as follows: T. F. MOSES to be Corresponding Secretary, and J. F. MEYER, Treasurer.

OBJECTS OF THE ASSOCIATION.

The objects for which this association was formed will be found stated in detail in the Constitution. For the better attainment of these the following *Sections* were adopted at the meeting of March, 1875, the members selecting for themselves the Section or Sections preferred for their field of labor:

1. Natural History and Geology.
2. Pioneer History and Archæology.
3. Anthropology.
4. Physics.

In order that the work of the Association might be, as far as possible, of general benefit to the community, another object, not specified in the Constitution, has always been had in view, the founding of a

PUBLIC MUSEUM.

In this museum the collections of the Association will be placed, and such other collections as shall be donated or loaned by individuals, whether members or otherwise. It is believed that all persons owning collections will recognize the unselfish character of the work in which

the Association is engaged, and that they will willingly place these
where they shall be accessible to every one and thus perform a larger
use than scattered private collections can possibly do. To further.
this object, efforts will be made to erect a suitable building at some
future day for the reception of the museum, and for other uses of a
similar character. Meanwhile, through the generosity of Mr. W. A.
BRAND, Postmaster at Urbana, the Association has been enabled to
carry out this plan to a certain extent, and has placed a part of its
collections in the Post Office building. Perhaps no location could be
selected more suitable for such a museum than the city of Urbana.
In the words of President GLOVER's inaugural address, "We have a
field that has scarcely been worked, and one that is replete with objects of interest and importance. Within a radius of a hundred miles,
lies a magnificent geological field with its paleontological treasures.
The drift and more recent deposits have been but little studied. In
natural history, zoology and botany, the region is a rich one; in ancient remains, the richest in America. Dr. FOSTER in his Prehistoric
Nations says that Ohio alone contains 10,000 tumuli or mounds, and
Mr. BALDWIN states that of these not over 500 have ever been
opened." The advantages to our city of such a collection of specimens of the natural history of the neighborhood as well as of the
relics of its pioneer inhabitants and of those races, unknown to history, whose numerous implements, ornaments and articles of domestic
use, annually turned up by the plow, make their mute appeal to our
human sympathies, are so obvious that they hardly need be stated.
Gathered together and preserved, they not only excite curiosity and
stimulate research on the part of every individual of the community,
both young and old, but they at the same time add to that vast mass
of material which is being accumulated all over the country, and out
of which is, at some day, to be evolved the history of our predecessors
upon this Continent. In so good a work will not every one aid, who
has in his possession a stone ax, arrow head, pestle, pipe, badge or
other relic, dug up from the soil or obtained from burial mounds, by
contributing them to the museum? Due credit will be given to
each depositor and all articles will be carefully labelled, and classified.
In the explorations that have been made by the Association, during
the past three years, much valuable material has been gathered together, especially of an archæological character, and when all the
necessary data are obtained it is proposed to publish a

MAP OF THE ANCIENT REMAINS OF MAD RIVER VALLEY.

This map will contain the exact location and dimensions of all the mounds and earth-works situated in the valley of Mad river and of its tributaries, and will constitute a valuable contribution to the general archæological map of Ohio. The importance of this work can hardly be over-estimated. Already many of the smaller earth-works have disappeared, having been destroyed through constant cultivation of the soil, and their location must be a matter of uncertainty and can be determined only by tradition. This process of destruction is going on with ever increasing rapidity and thoroughness under the demands of agriculture, so that in a few years it will be too late to secure the record of the site and extent of these ancient land marks. And yet there has been no period since their earliest discovery when so much interest has been aroused upon the subject of these remains, as at the present. The vast amount of material brought together at the Centennial Exposition and viewed by thousands from all sections of the country not only excited an intense interest in all matters pertaining to the prehistoric inhabitants of the continent, but it has also had the farther effect of spreading an intelligent appreciation among people every where of the character and variety of the relics found in the soil, and of the importance of collecting and preserving them. Of equal interest also are the earth-works and mounds, the only structures which have been left by time, as indestructible, when undisturbed by the hand of man, as the stone implements themselves. From these last must be gathered the greater part of all that can ever be known concerning the customs, domestic, ceremonial and religious, of their builders and of the degree of their civilization. The plan of the Association, in selecting a definite and limited field for its operations in this direction, will, it is believed, be productive of more valuable results than can be secured by desultory surveys and excavations in different parts of the State, and the work thus accomplished will supplement the labors of other similar societies. In the work of preparing the archæological map of Ohio the Association will cordially co-operate with these societies.

WHAT HAS BEEN DONE.

From the organization of the Association up to the present time the monthly meetings have been quite regularly held, and considering the small number constituting its membership, the meetings have been well attended. At these meetings discussions have been held

8

and papers have been read upon a variety of topics. The following synopsis, gathered from the minutes, will exhibit the range of subjects:

1. Inaugural Address, - - - President T. N. GLOVER.
2. Stone Bas Relief, found at Marblehead, O., - J. E. WERREN.
3. Method of Preserving Woods for Cabinet Specimens, L. C. HERRICK.
4. How to Geologize, - - - - T. N. GLOVER.
5. Shell Mounds on the Coast of Maine, - - T. F. MOSES.
6. Method of Preparing Casts of Specimens, - R. H. BOAL.
7. Contributions to a History of the Life and Times of Simon Kenton, - - - GEO. A. WEAVER.
8. The Drift of Champaign County, - - T. N. GLOVER.
9. Geology of Eastern Virginia, - - P. B. CABELL.
10. Geological Relations of Champaign Co., - T. F. MOSES.
11. Mineral Deposits of the Lake Superior Region illustrated with specimens gathered during the summer of 1875, - - - J. E. WERREN.
12. Report of Meteorological Observations made at Urbana, O., from the year 1852 to 1878, inclusive, - - - - - M. G. WILLIAMS.
13. Report of a Recent find of Human Bones near the Catawba Station of the C. C. C. & I. R. R., GEO. G. HARRIMAN.
14. Report of the finding of a Copper Needle and a deposit of 280 Flints. in Ashland county, O., also some bars of Lead used in Crawford's Campaign against the Indians, - - AARON ATEN.
15. Inaugural Address. - - President GEO. A. WEAVER.
16. Report of Rock Wells or "Pot Holes" examined at Eben-e-cook and Berlin Falls, Maine, and at Kanawha Falls, West Va., - - T. F. MOSES.
17. Report of Supposed Mound on the farm of Dr. Pearce, east of Urbana; also of the discovery of a grave of the daughter of the Chief of the Mingo tribe of Indians, at Mingo, Ohio, THOS. L. JOHNSON.
18. Report of the Roberts, Baldwin and other Mounds explored by the Association, - - T. F. MOSES.

9

Such of the above named papers as have been recommended by the Association for publication will be found in the present volume of Proceedings.

FIELD WORK.

This has been almost entirely confined to the examination of mounds, earth-works and aboriginal remains, and is detailed in full in the Report of the Ancient Remains of Mad River Valley. Nearly all of the members have engaged in the work, at one time or another, and have always found it a very enjoyable line of research. During the coming season it is proposed to make a systematic survey of the remaining earth-works in this neighborhoood, and to open such mounds as the time and means of the members may permit.

SPECIAL MEETINGS.

By invitation of P. R. BENNETT, Jr., the Association met at his house on the first Tuesday in December, 1875, where the members enjoyed the hospitality of Mr. and Mrs. BENNETT, and were also entertained by the exhibition of numerous microscopical objects prepared by Mr. BENNETT. The excellent microscopes of Dr. H. C. PEARCE and Dr. W. F. LEAHY, of Urbana, were kindly loaned for the occasion. It was decided to hold meetings of a similar character

from time to time upon invitation of members or citizens, the meetings to be of a more social character than the usual monthly meetings.
A special meeting was held in the afternoon of Dec. 19, 1876, which
was honored by the presence of E. S. Morse, of Salem, Mass.
Prof. Morse being requested to favor the Society with some remarks,
said:

REMARKS OF PROF. E. S. MORSE.

That he was pleased to see the Society in so flourishing a condition
as to membership and apparent enthusiasm; that the usual history of
scientific associations was an enthusiastic beginning, a gradual decline
and a final extinction, which in most cases arose *from a failure of the*
members to attend the regular meetings. Members should make it
a point of conscience to attend, if only to see that there was no business and to adjourn. He then gave a short history of a society of
which he had been a member which, by the continued persistence of
its supporters in all their duties, had attracted public attention to
such a degree as to secure legacies from unexpected quarters, and
though burned out twice, had arisen from its ashes and become a permanent and powerful institution. After some interesting general remarks on the importance and surpassing interest of scientific research,
he advised the Society to adopt some *special branch* of usefulness
and devote its energies to that, rather than to dissipate their efforts in
too wide a field. He suggested the importance and necessity of an
annual assessment for expenses, and that the usefulness of such an
association was by no means dependent upon the size of the town in
which it was located, but rather upon the enthusiasm and faithfulness
of its members.

A special meeting was also held on the evening of Jan. 3, 1878, to
inaugurate the new Society rooms in the Weaver building. Prof. E.
S. Morse, being in the city to deliver a public lecture, was invited
to be present, and he favored the Society with much interesting information about Japan, in which country he had become a resident.

PIONEERS' MEETING.

Agreeably to a resolution of the Society, a committee consisting of
Messrs. Young, Boal, Aten, Cabell, Moses and Harriman attended
the Pioneer meeting held at Mingo, in May, 1877. The visit was made
by invitation of the Pioneer Association of Champaign and Logan Co's,

and had for its object the presentation of the plan of work of the Society and its interests in general, with a view to securing the co-operation of the members of the former association. At the meeting addresses were made by Messrs. YOUNG and HARRIMAN on behalf of the Scientific Association, and Dr. MOSES gave an account of the recent opening of a mound. Many Indian and Mound Builders' relics, from the cabinet of the Association and that of Mr. ATEN, were exhibited with the hope of exciting an interest in the collection of these objects, and it was suggested that all having such articles in their possession should deposit them in the museum of the Association for their better preservation, until a building suitable for the public museum can be secured. This suggestion was quite generously responded to on the part of several persons. The results of this meeting were so satisfactory, that it was thought advisable to send delegates to future meetings, and, if possible, secure a more intimate connection between the two Associations for the more successful pursuit of such objects as they have in common. It is important that our Pioneers, who are passing away, should leave behind them, not only the traditions of their early struggles and triumphs, but a more substantial memorial in the way of historical records and actual mementoes of their mechanical and domestic appliances. To gather these mementoes together and place them in a museum where they may always be seen by those who have inherited the land secured with so much difficulty from the savage, and thus to preserve a lively remembrance of the early settlers of the county among their descendants, is one of the aims of the Scientific Association.

NEW SOCIETY ROOM.

In January, 1878, the Association took formal possession of the new room in the Weaver building which had been especially fitted up for its use under the superintendence of Mr. GEO. A. WEAVER. Up to this time the meetings had been held in the rooms of Dr. R. H. BOAL, who had kindly granted the use of them free of charge. Although the present room is well adapted for the holding of meetings and as a place of deposit for a certain class of the Society's collections, it is probable that at no distant date more space than it affords will be required, unless the plan of the public museum can be carried out.

12

INCORPORATION.

At the meeting of February, 1878, a resolution was passed to the effect that it is the desire of the Association to become an incorporated body, and the executive committee was directed to take steps to secure the necessary articles of incorporation. Six Trustees were also appointed to serve for periods of one, two and three years. The Articles of Incorporation were duly executed and filed May 10, 1878.

OFFICERS ELECTED Oct. 19, 1874.

President—T. N. GLOVER;   Vice President—P. R. BENNETT, Jr.
Recording Sec'y—W. F. LEAHY;   Coresponding Sec'y--T. F. MOSES
Treasurer—J. F. MEYER;   Curator--R. H. BOAL.

OFFICERS ELECTED Nov. 16, 1875.

President—T. N. GLOVER;   Vice President P. R. BENNETT, Jr.
Recording Secretary—R. H. BOAL;   Cor. Sec'y—T. F. MOSES;
Treasurer—L. C. HERRICK;   Curator—R H. BOAL.

At the meeting of Feb. 15th, 1876, Mr. GLOVER tendered his resignation as President, being about to change his residence to Joliet, Ills. At the following meeting, T. F. MOSES was elected to fill the vacancy.

OFFICERS ELECTED Nov. 26, 1876.

President—GEO. G. HARRIMAN;   Vice President—P. R. BENNETT, Jr.;
Recording Sec'y—P. B. CABELL;   Cor. Sec'y—T. F. MOSES;
Treasurer—C. G. SMITH;   Curator—R. H. BOAL.

OFFCERS ELECTED OCTOBER 17, 1877.

President—GEO. A. WEAVER;   Vice Pres.—HAMILTON RING;
Recording Sec'y—J. S. PARKER;   Corresponding Sec'y—T. F. MOSES;
Treasurer—AARON ATEN;   Curator—R. H. BOAL.

TRUSTEES ELECTED FEB. 19, 1878.

FOR THREE YEARS.

GEORGE A. WEAVER,   R. H. BOAL.

FOR TWO YEARS.

JNO. H. YOUNG,   THOS. F. MOSES.

FOR ONE YEAR.

CHAS. G. SMITH,   HAMILTON RING.

CONSTITUTION.

ARTICLE 1. The Society shall be called The Central Ohio Scientific Association.

OBJECTS.

ART. 2. The Objects of the Association shall be,

First—The cultivation of Physical and Historical Science.

Second—The study of the region around us and its inhabitants.

Third—The development of a scientific taste in the community.

Fourth—Mutual acquaintance among scientific workers.

MEMBERSHIP.

ART 3. All members shall be chosen by ballot, after having been nominated at a preceeding meeting. The affirmative votes of three-fourths of the members present shall be necessary to a choice. Corresponding, Honorary and Associate members may be elected, and shall be free from all fees and assessments. Ladies only shall be eligible to the class of Associate members. Active members only shall be entitled to vote.

ASSESSMENTS.

ART. 4. Any person, on being elected to the Association, shall pay an initiation fee of two dollars ($2.00), and such person shall not be considered a member or entitled to the privileges of membership until his initiation fee is paid. The annual assessment for each member shall not exceed Ten dollars ($10.00), and shall be made at or between the October and January meetings of each year.

OFFICERS.

ART. 5. The officers of this association shall consist of a President, Vice President, Recording Secretary, Corresponding Secretary, Treasurer, and Curator. They shall hold their offices one year, or until their successors shall be appointed. They shall be elected by ballot.

EXECUTIVE BOARD.

ART. 6. The President, Recording Secretary, and Treasurer, shall constitute an Executive Board.

ART. 7. By-laws for the more particular regulation of the Society may be made.

AMENDMENTS.

ART. 8. Amendments to or changes in this Constitution may be made by a two-thirds vote of the Society at any regular meeting, due notice having been given at a previous meeting in writing, signed by three members of the Society.

BY-LAWS.

Section 1. The Annual Meeting shall be held on the third Tuesday in October, when the election of officers shall take place.

Sec. 2. The Regular Meetings of the Association shall be held on the third Tuesday in each month.

Sec. 3. Any member who shall fail to pay his assessment on or before the first day of April, after having been duly notified, shall forfeit his membership.

Sec. 4. The Order of Business shall be,

First—Calling of the Roll and reading of the minutes of the previous meeting.

Second—Report of the Treasurer and Payment of Dues.

Third—Reports of Committees and Correspondence.

Fourth—Reading of Communications and General Business.

Fifth—Adjournment.

List of Members.

ACTIVE MEMBERS.

Theo. N. Glover, Dowagiac, Mich.

L. C. Herrick, Woodstock.

P. R Bennett, Jr., Urbana.

Thos. F. Moses, Urbana.

R. H. Boal, Urbana.

Wm. F. Leahy, Urbana.

Geo. A. Weaver, Urbana.

J. F. Meyer, Urbana.

Jno. H. Young, Urbana.

Thos. L. Johnson, Mingo.

Jeremiah Deuel, Urbana.

J. F. Gowey, Urbana.

P. B. Cabell, Urbana.

J. E. Werren, Urbana.

Hamilton Ring, Urbana.

D. H. Sherman, Urbana.

J. S. Parker, Urbana.

C. G. Smith, Urbana.

Geo. G. Harriman, Urbana.

Aaron Aten, Urbana.

S. F. Woodard, Osborn.

Jas. Pillars, Lima.

CORRESPONDING MEMBERS.

David T. Robinson, Urbana, Ohio.

Prof. D. S. Jordan, Pardee University, Indiana.

Dr. R. M. Byrnes, Cincinnati, Ohio.

Lyman C. Draper, Wisconsin.

Prof. Edw. S. Morse, University of Tokio, Japan.

HONORARY MEMBERS.

Milo G. Williams, Urbana.   John. H. Klippart, Columbus.

Donations to the Cabinet

From 1874 to 1878.

R. H. Boal—Copy of Boston Gazette and Country Journal, of March 12, 1770.

Jno. H. Klippart, Columbus, O.—Ohio Agricultural Reports for 1859, 1871, 1872.

D. T. Robinson—Niagara Fossils from Madison, Indiana.

Maryland Academy of Sciences—Charter, Constitution and By-Laws.

Geo. A. Leakin—Pamphlet entitled The Periodic Law. By G. A. Leakin, Baltimore, Maryland.

Smithsonian Institution, Washington, D. C.—Miscellaneous Collections, Nos. 34, 160, 261 and 278.

Peabody Academy of Sciences, Salem, Mass.—Annual Reports from the First to the Tenth, inclusive.

F. W. Putnam, Salem, Mass.—Remarks on the Family Nemophidae. Notes on Liparis and Cyclopterus. Mounds at Merom and Hutsonville. Notes on Ophidiidae and Fierasferidae. Description of stone knives found in Essex county, Massachusetts.

Interior Department, Washington, D. C.—Synopsis of Acrididae of North America, Thomas; Extinct Vertebrate Fauna, Leidy; Bulletin of U. S. Geological and Geographical Survey of the Territories, Second Series, No. 1.

Report on Cretaceous Flora, Lesquereaux.

U. S. Geological Survey of the Territories, by F. V. Hayden. Reports of 1867, 1868, 1869, 1871, 1872 and 1873.

Flora of Colorado—T. C. Porter and J. M. Coulter.

List of Elevations, Third Edition, by Henry Gannett.

Birds of the North-west—Coues.

Department of Agriculture, Washington, D. C.—Reports of 1872 1873.

Hon. T. A. Cowgill, Ohio House of Representatives—Geology of Ohio, Vol. 2. Ohio Statistics for 1875.

Griffith Ellis, Esq.—Palæontology of Ohio, Vol. 2.

Hon. W. R. Warnock, Urbana—Ohio Centennial Report.

Robert Clarke, Esq., Cincinnati, Ohio—Prehistoric Remains, Cincinnati, Ohio, R. Clarke.

S. W. Garman, Cambridge, Mass.—The Skates of the Eastern Coast of United States. S. W. Garman.

Davenport Academy of Natural Science, Davenport, Iowa—13 Photographs of Pipes, Copper Axes and Skulls.

Cincinnati Society of Natural History—Proceedings, Number 1.

M. G. WILLIAMS, Esq.—Report of his Meteorological observations made at Urbana, O., from 1852 to 1868.

JOS. CHAMBERLAIN, Rush township, Champaign county, Ohio—Specimen of Bog Iron Ore.

T. N. GLOVER, Joliet, Illinois—Cast of Stone Image Pipe found on banks of the Darby River, Pleasant Valley, Ohio.

JNO T. HUNTER, Mingo, Ohio—Indian Tomahawk, of British make, found in 1828 near Mingo.

JAMES McLARY, Mingo, Ohio—Comb Saw brought to America from Germany, by Jno. Barrett, in 1806.

DR. L. C. HERRICK, Woodstock, Ohio—Indian Skull dug up in Wayne township, from a gravel bed.

CHICAGO ACADEMY OF SCIENCE, through C. G. Smith, Esq., of Urbana, Ohio— Collection of 76 species Shells from coast of Florida, Gulf of Mexico and California.

T. F. MOSES—Ancient Vase from Drevant, France; Arrow Head from Champaign county; Minerals from Buchanan's Farm, Lancaster, Pa.

AARON ATEN—Stone Ax, Wyandot county, Ohio; 11 Arrow Heads from a lot of 182 found buried 12 inches below surface of ground, Wyandot county, Ohio; fragment of ancient mill-stone, Lead Ingot, Ashland county, Ohio; Cannon ball from battle-field of Shiloh; Concretions and water-worn stones from Wyandot county, Ohio; Mortar, Pestle, and 3 Fleshers, Clay Pipe, numerous specimens of Galena, Iron Ore and Quartz, from Washington county, Miss.

T. N. GLOVER—Ancient Pottery from Darby Plains, Stone Pipe, fragments of Elk Horns found in peat bog near Woodstock, Ohio; Corniferous Fossil, Coal Plant.

W. R. HARDMAN, Enon, Ohio—Stone Flesher, Stone Hammer and two Flints.

W. K. PATRICK, Urbana, Ohio—2 Celts, 2 Axes, Stone Hammer, Gorget, Sinker, 3 Flints, 1 Pipe from Licking county, O.

WM. PATRICK, Urbana, Ohio—Part of Simon Kenton's coffin.

P. R. BENNETT, Jr.—Stone Ax.

JNO. H. YOUNG—Pestle, 2 Flint Implements and portion of mill-stone plowed up on his farm in Urbana.

E. E. McFARLAND—Mica from a mound on the J. D. Wilson farm.

LAUSON SHOWERS, Urbana, Ohio—3 Flint Arrow Heads.

FRANK MAGREW, Urbana, Ohio—Flint Arrow Head.

M. ARROWSMITH—3 models in wood of ancient badges from a mound on the farm of the late Ezekiel Arrowsmith, of Madriver township, Champaign county.

DR. H. C. HOUSTON, Urbana, Ohio—2 Terra Cotta Images from Mexico.

DR. W. F. LEAHY—Lead bullet found in the heart of a tree from Saginaw.

GEO. A. WEAVER—Specimens of Native Woods.

CHAS. VAN METER—Arrow Head from near Circleville, O.

DR. H. RING- Stone Pestle from the Dallas farm, below Urbana, also piece of Conglomerate.

R. M. GWYNNE—Set of Minerals, Fossils and Fresh Water Shells.

EDGAR HODGE, Champaign county—Set of 45 Flints from S. E. part of county, near Buck Creek.

FRANK BOAL, Urbana, Ohio—Specimens of Silurian fossils from Warren county, O.

JAS. PILLARS, Lima, O.—Sea Mat.

CLAY JOHNSON, Urbana, Ohio—One Stone Pipe found in Champaign county, O., (on deposit.)

DR. H. C. HOUSTON—Sea Bean from Florida.

W. P. DABBS, Urbana, Ohio—Wood showing work of tree-borers.

JAS. LANDIS, Urbana, Ohio—Cotton Bole, Jackson, Tenn. Snout of Saw Fish.

MRS. MAJOR HUNT, Clarke county—String of Wampum from an Indian Tribe in the North-west.

CHAS. ROBERTS, Clarke county—5 Pestles, 1 Hammer, 5 Mauls, 2 Axes and 6 Flints from Clarke county, near Buck Creek.

AARON ATEN--Part of rib of mastodon, petrified; Fossil tooth, 3 pairs of Deer horns, Tomahawk. Pipe, from Crawford's battle field.

PROF. J. E. WERREN, Urbana University--Original chart of earth-works on Haddix Hill, Osborn, O. Also, original drawings of Plates 9 to 12, inclusive, of the Heliotype illustrations of the Proceedings.

D. H. SHERMAN, Urbana, Ohio—Original Drawings from which the Heliotype illustrations of the Proceedings were made, Plates 1 to 8 inclusive.

CHAS. G. SMITH, Urbana, Ohio—Pair of large Elk horns.

M. H. CRANE—Rattlesnake killed near Mad river, Champaign county, Ohio.

D. H. SHERMAN—United States Geological Survey. Featherstonhaugh, 1836.

MR. HADDIX, Osborne, O.—Stone Pipe from Mound.

DR. L. C. HERRICK, Woodstock, Ohio—Wool-wheel, Flax-wheel, Reel and Hackle. These were formerly owned by Mrs. Mary Overfield, of Rush township, Champaign county, O., and were brought by her from Virginia.

INAUGURAL ADDRESS

—OF—

PRESIDENT T. N. GLOVER,

Delivered November 17th, 1878.

Gentlemen of the Association :

When the President of such a society as this enters upon the discharge of his duties, it is customary for him to deliver an address. I follow the custom; but, instead of talking upon a scientific subject as is generally done, I confine my thoughts to the work of our Association.

Our constitution declares that the objects of our meetings shall be:

1. The cultivation of Physical and Historical Science.
2. The study of the country surrounding us, and its inhabitants.
3. The cultivation of a scientific taste in the community.
4. The mutual acquaintance of scientific students.

I suppose the most prominent thought one has, when he hears the first object stated, is the question, Why are the two classes of science, Physical Science and Historical Science, connected? It is not usual for such societies as this to aim at more than one class; and, in this age of specialties, if we would do good work, we must confine our attention within as narrow limits as possible.

I would answer that the two are connected in this society, because by nature they are connected. No one knows a subject when he is acquainted with it in its present phase only. Since scientists have studied the antiquity of the human race, their views of ethnology have been materially changed. The study of paleo-botany has changed some old views of modern botany. That man has an influence on the earth is admitted; and, in order to trace this influence, we must have the aid of historical sciences. A writer in the Encyc. Brit. groups the mathematical and physical sciences, claiming that with the exception of magnetism and electricity all the so-called physical sciences have been aided largely by the mathematical. Though it may be submitted whether or not the mathematical sciences do not constitute a division of the physical, yet on the principle he has laid down,—the aid the one has given the other,—I claim that the historical and physical sciences must not be separated.

And in the question which may here arise, Which sciences constitute the physical sciences? a grave difficulty presents itself. We have no standard of definition. One author, (another writer in the Encyc. Brit.), says all the physical sciences can be grouped under Geology and Astronomy; the author quoted above says they are only electricity and magnetism; while if I remember aright, a much quoted author, Dr. Campbell, in his Philosophy of Rhetoric, follows the old Greek classification and includes Psychology and Natural Theology.

The question is one of importance to us, since we are meeting on common ground and wish to avoid all questions which tend to excite dissensions. Hence I take the liberty to recommend earnestly that you consider this matter, and as soon as possible form a list of those sciences which you deem physical, and which shall be allowed on the floor of this Association.

In regard to the historical sciences, the same trouble, in a measure, is encountered; and the recommendation just made may be applicable. Yet the way is more clearly defined. Prof. Neander has given a beautiful definition of history. "Its office," says he, "is to impart unity to the consciousness of mankind when it has been divided by Time. It originates in the effort to connect the present with the past." So whatever bears directly upon this is Historical; be it in point of age, a thing of a thousand or ten thousand years ago, or of scarcely five minutes past; be it in point of matter, a conclusion of Comparative Philology. a conclusion derived from comparing manners and customs and beliefs of people; an examination of the remains of ancient cities, of the grave mounds of Europe. Asia, or America; of cave remains; a transcript of the scrolls of Egypt or India; copies of records from any part of the globe at any period of time and of any nature whatever; or a statement of what men now think, how they dress, what they believe, and how they live.

The second object of our Association is the study of the surrounding region and its inhabitants. Here we have use for both the Physical and Historical sciences. Here we have a field which has never been worked, and which is replete with objects of interest and importance. Within a radius of an hundred miles lies a magnificent geological field rich in paleontological treasures, beautiful for historical arrangement, and worth an investigation by the economic geologist. The drift, and more recent deposits of America, have never yet been worked up. In Natural History, that is, in zoology and botany, our field must not be neglected. In Indian remains, it is very rich ; and, in prehistoric remains, the richest in America. Dr. Foster, in his Prehistoric Nations, says that Ohio alone contains over 10.000 tumuli, and Mr. Baldwin states that not over 500 of these have been opened.

Then we have customs and beliefs and methods of talk which are worthy of preservation. A man not long ago gave me his reasons for believing that boulders grow after the manner of vegetation ; the idea is worth preserving as a matter of history. In the matter of provincialisms, we can find much profit. Then in working up the early settlement of this country by the whites—the manners and customs of the Indians who lived here at that time—the field is comparatively untouched.

In regard to the third object of our Association, there is pleasure in knowledge, and scientific knowledge is much easier gained than many other kinds. So when men pass the matter carelessly by they deprive themselves of much pleasure. But more. Science itself is a loser thereby. To appreciate a thing we must understand it. Men appreciate money, for they know its use. Cortez destroyed the Mexican remains because he did not know their nature. The relics and the specimens of natural history in this country are thrust aside uncared for. Corals of the most common varieties are showered upon the collector under various names, such as petrified deers' horns, honeycombs, &c., while a really valuable coral, or a stone ax. or arrow head is passed by. A farmer has in his field a prehistoric mound—he either plows it down without any care, or if he examines it, he does it with so little care that his work is almost worthless. More than this, in an economic sense, science has claims—for a people who appreciate its surroundings are always wealthier than one which does not.

In regard to the mutual acquaintance of students: Of the many points of interest which our society presents, I esteem no one of more importance than this. Troubles present themselves to one student and keep shy of another. One student can benefit another by exchanging with him observations and speculations, by pointing out his errors, by discussing scientific views. One potent cause of men's ignorance of every-day subjects is, they know not how to go to work at them.

This much, gentlemen, as to the objects of our organization. We do not meet here as a social school, as some people have supposed. We do not come here to do our work, but to compare our works, to criticise them, to aid each other in our difficulties, to preserve whatever we may have gathered. And this leads me to the one thing which we must bear in mind: That we are to work. Our meetings are not to be playspells; nor are we to work carelessly. First-class work is the demand—work done carefully, conscientiously. Treat our subjects thoroughly. We have a superabundance of science for popular use; altogether too much scientific work, so-called, done when one has wandered into the fields and seated himself on the velvet moss beneath the leafy canopy of some umbrageous oak; too many scientific Fourth of July orations. The need is of accurate, complete work, independent work—reliability must characterize it. We must be an authority in our field.

REPORT OF THE ANTIQUITIES OF MAD RIVER VALLEY.

BY PROF. THOS. F. MOSES, URBANA UNIVERSITY.

The relation of the physical features of a country to the distribution of its population is a well-known fact, especially as regards the location of cities and towns along the shores of seas and lakes and upon the banks of navigable rivers. At the present day, since the general introduction of railroads and canals, centers of population are found somewhat remote from the old highways of commerce, so that this natural relation is somewhat obscured. In the prehistoric times of our continent, however, when the streams were navigable in their smallest branches, and almost to their very sources, by the vessels of light draft common to primitive peoples, we should expect to find a more uniform relationship between the early settlements and the water-courses than now exists; and it is the fact that we do find the mounds and earthworks constructed by the early inhabitants in close proximity to the rivers and their tributaries, those of less extent being found along the smaller streams while those of greater magnitude occupy the larger river valleys and positions of greatest natural advantage. Hence, in describing the ancient remains of the valley of Mad river we follow a natural method and confine our attention to a well-defined area, the study of which must necessarily yield more satisfactory results than that of one bounded by county lines or other arbitrary limits.

As at the present day the topography of a country in other respects has greatly to do with the location of the settlements, so in these times the sites for erecting structures were carefully selected upon the elevated plateaus and alluvial terraces where the salubrity of the air and the fertility of the soil afford peculiar advantages. Frequently the structures occupy the summits of lofty hills overlooking the valleys, and it is noticeable that the situations thus chosen are remarkable for

the beauty of prospect, showing that this also had its influence as well
as the more utilitarian objects mentioned above. These latter structures were evidently chosen as burial places or as being suitable for
purposes of observation and defense. Those occupying the alluvial
plains are usually more extensive in character and were undoubtedly
the seats of large towns. So well was their site chosen with reference
to the natural advantages of the country that most of them are
to-day covered with flourishing cities. In the words of Mr. E. G.
Squier :

"It is worthy of remark that the sites selected for settlements, towns
and cities, by the invading Europeans, are often those which were the
especial favorites of the mound-builders, and the seats of their heaviest
population. Marietta, Newark, Portsmouth, Chillicothe, Circleville and
Cincinnati, in Ohio; Frankfort in Kentucky, and St. Louis in Missouri,
may be mentioned in confirmation of this remark. The centers of population are now where they were at the period when the mysterious race
of the mounds flourished."*

GEOLOGY.

The valley of Mad river, the chief tributary of the Great Miami,
presents a well-marked geographical region characterized by numerous
interesting topographical features. The main stream takes its rise but
a short distance south of the ridge which constitutes the *divide* between the waters flowing north to Lake Erie and those flowing
south to the Ohio, the upper part of the Scioto valley alone intervening. The head of the stream lies east of Bellefontaine, in
Logan county, in the Huron shale where, according to President
Orton's report in the State Geological Survey (vol. I., page 454),
the source has an altitude of 1,438 feet above tide water, being an
elevation equal to that of any other in the State. In the upper
part of its course the bed of the river lies over the Corniferous
limestone and during its passage through the northeast part of Champaign county it is underlain by the Helderberg limestone. For the
remainder of its course in Champaign county the river meanders
through a series of peat bogs and marshes which overlie the drift. In
this part of its course it is fed by numerous perennial springs which
give to the stream a permanent character even in the dryest seasons.
In the neighborhood of Urbana these springs are very numerous and
the supply of water they afford is practically inexhaustable. The
construction of the well for the Water-Works recently erected at Urbana affords the following facts bearing upon this point: The well is

*Ancient Monuments of the Mississippi Valley.

dug in the bottom just at the foot of the terrace upon which the city stands, not far from the junction of the Pennsylvania and Mad river railroad. The depth of the well below the surface of the water is nine feet and its diameter fifteen feet. In its construction the utmost difficulty was encountered owing to the rapid inflow of water, two rotary pumps and three steam engines being employed to keep it clear of water while the excavation was going on. Mr. Parrott, the engineer of the Water-Works, informs me that in making the "duty test," 2,000,000 gallons were thrown out in twenty-four hours. After pumping two hours, the level of the water had been lowered three feet, but there was no farther sinking during the remainder of the test.

The whole region of Mad river valley from the line where the Helderberg formation breaks off in the north-eastern part of the county to the first contact of the river with its rocky borders is, in fact, a vast basin occupied in many places with vegetable accumulations saturated with moisture. The depth of these accumulations to the underlying gravel is not many feet, and the width of the basin varies from one-half to three miles. The river itself occupies but an insignificant part of the basin except in seasons of heavy rains, when it runs riot over the whole plain, the sheet of water becoming in some places several miles in width. It is partly to this erratic tendency and partly to its tortuous channel and rapid flow that the river is commonly supposed to owe its name, but I am informed by John H. James, Esq., of Urbana, that its original name was from the Indian tongue, the translation of which is "River of Anger."*

The hills which here form the borders of the valley are entirely composed of drift materials, no rock being anywhere exposed, even in the bed of the stream. Indeed, supposing a uniform dip to characterize the underlying strata, similar to that exhibited at Springfield, the nearest rock or upper stratum of the Niagara limestone would be 85 feet below the surface at the railroad station at Urbana. This theoretical estimate, made by the writer from a comparison of the A. & G. W. railroad profiles with the observed dip of the strata, can only

*This information was communicated to Col. James by Robert Flemming who, when a boy, had been the servant of one McKee, an Indian trader. Before the revolution McKee resided at Fort Pitt, but on the Declaration of Independence he cast his lot with the British and went to reside with the Indians beyond the Ohio, taking the lad Flemming with him. Flemming spent many years with the Indians, learning their language and customs. He stated that the name "River of Anger" was given to this stream because the Indians here brought their captives to be burned. The place where this punishment was executed was near the source of the river in Logan county, on a tract of land formerly belonging to Col. James.

be verified by a series of borings. In this part of its course Mad river receives a number of tributaries from either side, the most considerable on the east being Macachack, Kingscreek and Pleasant-run, and on the west Glady's creek and Muddy creek, which drain the high land near Spring Hills, and, proceeding south, Spring, Nettle and Stormes creeks, and Black Snake run, on the borders of Clarke county.*

The country drained by these tributaries is mostly of a gently undulating character, the knolls being composed of clay, sand and gravel, the materials of which have been largely furnished by the rocks of Ohio. There are also a number of well-marked terraces where the gravels are overspread with a rich alluvial soil, either with or without an intervening stratum of clay. Along the valley of Buck creek, in the eastern part of Clarke county, the gravel hills slope gently down to the level of the broad plain in which the stream lies, presenting a very picturesque appearance. In the opinion of President Orton these ridges are not the remains of a general superficial deposit, but were laid down during the period of the post-glacial in the same form that they now possess.*

The shape and general appearance of the ridges certainly favor such a supposition. Their gentle and uniform slope, the numerous gorges which penetrate their sides giving to them a peninsula-like form but having apparently no present connection with the drainage system, the occasional depressions found upon their summits, all these conditions seem to favor the notion that they were deposited in strong counter currents of water very much as we see gravel banks forming along the course of our rivers at the present day. The examination, however, which the writer has made of sections of several of these gravel hills in the neighborhood of Catawba station, on the C. C. C. & I. R. R., and at Baldwin's mill, do not seem to corroborate this theory. In these sections the horizontal lines of stratification are clearly seen, and in every instance the strata run quite to the edge of the hill where they terminate abruptly in the thin surface covering of the soil. These ridges are the favorite sites selected by the Mound-builders for their struct-

*I am informed by Col. J. H. James that the present Dugan's run, emptying into Mad river south-west of Urbana, is an artificial tributary resulting from the construction of a drain from a marshy tract of land known as Dugan's prairie. In constructing the ditch it was necessary to cut through a ridge, and the natural drainage was formerly, probably toward Buck creek.

†State Geological Survey, Vol. 1, p. 459.

ures in Champaign and Clarke counties, and they are also frequently found to contain large deposits of human remains over which no structure has been raised. The origin of these deposits is as much involved in obscurity as that of the mounds and their contents.

That portion of Mad river valley which lies in Clarke, Greene and Montgomery counties has been so well described in President Orton's reports contained in the "Ohio Geological Survey," that it would be unnecessary to attempt the description here, except for the purpose of presenting the subject as a whole and in a connected manner, hence a brief summary only will be given. In the upper half of Clarke county the topography is essentially the same as in Champaign county, the only marked change observable being that the valley is generally narrower and the bordering hills higher. These side hills, like those bordering the Ohio at Cincinnati, are in reality only the margins of the general level of the country, and mark the limits of a former period of erosion. In the latter instance the erosion has been effected principally through the Silurian limestone, while in the region we are now describing it has taken place in the overlying drift deposits. The resemblance to hills is increased by the lateral channels of erosion which communicate with the main valley. Most of these communicating channels are short, steep gorges rising rapidly to the general level of the plateau, while others traverse a large extent of country and contain the present tributaries of the river. In this part of its course the river traverses a large "cedar swamp," in which grows the white cedar or *Cupressus thyoides*. This swamp and a smaller one farther up the stream in Champaign county, are the only localities in this part of the State where these trees are found native. In the peat bogs, especially those bordering Buck Creek, large trunks of the red cedar are often met with several feet below the surface, but none are found growing anywhere in the vicinity at the present day.

If the surface deposits of Champaign, Clarke, Greene and Montgomery counties were stripped off and the rocky floor laid bare, we should see a succession of limestone steps or platforms rising from the south and ascending toward the north-east of this whole region. Across the northeastern part of Champaign and Clarke counties would be seen the upper platform of this flight of steps composed of the Helderberg limestone. Next in order would be seen a broad platform of Niagara rock, spreading over, probably, the whole of Champaign and the greater part of Clarke counties.

Bordering this platform in the south-western part of Clarke county would be seen the next step in the series, that composed of the Clinton limestone, much lower and narrower than the others, and of very irregular outline. Finally, spread over the rest of the region indicated, would appear a floor of Blue limestone, hidden from view here and there by patches or islands of the layers already mentioned, some of these islands being composed of the remains of the next higher step, some of the two preceding ones, but none showing any traces of the highest or Helderberg formation. The whole surface would appear scarred and seamed by erosive agencies, some of which are now in operation and some of which have long ceased to exist. Let us now suppose this series of steps to be inclined somewhat toward the north-east so that their outer edges rise to nearly the same level and there inner angles form a series of shallow troughs. The long axis of these troughs extends generally from the north-west to south-east and Mad river traverses them nearly at right angles, its general trend being from north-east to south-west. The edge of the Helderberg step is crossed in the northern part of Champaign county as we have already seen. After this the river is bedded in the overlying drift deposits with which the whole trough at the foot of the Helderberg layer is filled. Of this basin and its marshy character we have already treated. The next step reaches the surface in the neighborhood of Springfield, that of the Niagara limestone. Through the edge of this the river has worn a narrow gorge for some distance. The picturesque cliffs at Moore's and Holcomb's quarries, nearly one hundred feet in height, are good examples of this erosion. Here the river valley is suddenly narrowed, its width opposite Moore's quarries being not more than one-quarter of a mile. Before this rocky barrier was worn away the river may have had a considerable fall at this place, the rock structure being precisely like that at Niagara Falls, and continuous with it. We may also suppose the marshy region behind to have been occupied by a lake or series of lakes, before the present free outlet for the waters was afforded. At Springfied, Buck creek enters Mad river from the east, and two miles below the city Mill creek. These latter have also wrought the latter part of their channels through the Niagara limestone. Five miles below Springfield, at Snyder's station, the river traverses the next layer of the series, the Clinton limestone. The valley now rapidly broadens and the course of the river lies over the blue limestone of the Cincinnati group until it debouches into the Great Miami at Dayton. At Osborn it receives its last principal tributary, that of Mud run, the southern boundary of which is formed by the Clinton limestone. Haddix Hill, on which is an extensive system of earth-works constructed by the Mound-builders is one of the islands or "outliers" of Clinton limestone left by some former erosive action of great magnitude. Other detached masses of this limestone are found in the neighborhood. The cause of this extensive erosion is usually ascribed to glacial action of which there are abundant evidences. The polished and striated surfaces of the cap-rock of the Helderberg at McComsey's quarry, east of Urbana, in Champaign county, of the Niagara at the Springfield and Moore's quarries, of the Clinton limestone at Snyder's Station in Clarke county, show that the great ice-mass ground its slow way along over the edges of all the steps in the series we have described. A very remarkable example of this glacial polish has been brought to light in stripping the surface of Booher's quarry at Taylorsville, in Montgomery county. This quarry is situated on the summit of a large island of Clinton and Niagara limestone, which is bounded on the west by the valley of the Great Miami, and is separated from the same formations on the east by an ancient eroded channel which, in the opinion of President Orton, was the former bed of this stream This channel is now occupied by Honey creek, flowing north-west and emptying into the Miami, and by a small tributary of Mad river which takes a southerly direction, reaching the river at Osborn. In opening this quarry several feet of clayey soil were removed to expose the rock which at this place is the lowest member of the Niagara series, the Dayton limestone. The whole rocky floor thus exposed, already several acres in extent, is smooth and polished and planed down almost to a perfect level. It is cut into tables for flag stones and similar uses, blocks ten feet square being sometimes removed which present so smooth and true a surface that they seem to have been artificially dressed. The cyathophylloid corals and other fossils which occur in the stone are finely exhibited in vertical and cross sections. Comparatively but a small portion of this polished floor has as yet been uncovered and its probable extent has been estimated at not less than 200 acres.

ANTIQUITIES.

The location of the earth-works of Mad river valley that have thus far been examined are found to bear a close relation to the topographical features of this region as described above. They occur usually, on the high land overlooking the river valley, the exceptions to this rule being nearly always in the upper part of its course where a mound is occasionally found located on low ground at the junction of the main stream with one of its smaller tributaries. The works consist almost entirely of mounds, there being but one enclosure, as far as is known at present, namely, that situated on Haddix Hill near Osborn, and described in Prof. Werren's paper.* The mounds vary greatly in size. The smaller ones are usually low and flat on the summit, and might be styled disk-shaped mounds. These are from three to five feet in height and from thirty to fifty feet in diameter. Another class of mounds is more conical in shape, varying from eight to fifteen feet in altitude and having a diameter at the base of from seventy to eighty feet. The very large mound at Enon, in Clarke county, belongs to the class of Great Mounds, like those at Miamisburg, O., and Gravecreek, Va., described in Squier and Davis' Monuments of the Mississippi Valley. This mound is from fifty to sixty feet in height and is situated a short distance from the village of Enon, upon the east bank of the river. The internal structure of all the mounds of this region that have been opened is nearly homogeneous in character, being generally of a clayey loam like the surface soil. In regard to the relation of the mounds to each other, sufficient data have not yet been obtained upon which to base a definite statement, but it is observed that in the case of those situated upon high ground, one or more may be distinctly seen from the summit of another, as in the case of the Baldwin and Roberts mounds on Buck creek, and of the mounds at Enon, Haddix Hill and Kauffman's farm, suggesting the idea that they may have been used as signal stations. Of the mounds thus far opened by the Association, those upon the farms of Mrs. Samuel Baldwin and of Mr. Charles Roberts have yielded the most important results, and the description of their excavation and contents will be given in detail. The Baldwin mound was opened in the summer of 1876 and the Roberts mound a year later, but an account of the latter will be given first.

*Another earth-work is reported about five miles east of Springfield, but it has not yet been examined.

THE ROBERTS MOUND.

This Mound, like the Baldwin Mound, is located upon a high hill composed of drift gravels and sand, the materials having been chiefly derived from the limestone strata of our own State. Standing upon its summit a wide and beautiful prospect meets the eye in whatever direction one may turn. On the east the horizon is bounded by a range of hills. These hills are in reality the termination of a broad plateau and they indicate the contour lines of the eroded valley formed during some former geological period, into which valley now flow the streams which furnish the natural drainage of the country. Similar plateaus stretch away to the north and to the south and their numerous upland farms, teeming with abundant harvests, betoken the extraordinary fertility of the soil. One of these elevated plains is styled "Pretty Prairie." This name is applied to the southern part only of the northern plateau, but geologically it extends to the eastern side of Urbana, the city itself being placed upon a lower terrace, and is the whole tract included between the valley of Madriver and that of its eastern tributary, Buck Creek. But perhaps the most remarkable feature of the landscape is the broad valley itself, which sweeps down between the plains above described from near Mechanicsburg on the north-east, and taking a course due west as it flows by the base of the hill upon which the mound stands, finally trends away to the south-west, broadening as it goes, and, at last, lost to view in the distant horizon. One cannot fail to be impressed with the idea that this valley once held a noble river, smoothly flowing through its generous channel and hiding from view the present alluvial plain. The only remnant of this river, if such there were, is the little stream called Buck Creek, so named from the manner in which its smaller branches here unite with the main trunk like the antlers of a stag. Standing upon this mound, the prehistoric inhabitant could see the mound on what is now the Baldwin farm, crowning the summit of the opposite bank, and give to his friends across the stream a token of welcome or a signal of approaching danger. These sites, selected as they undoubtedly were with unusual care, as burial places for their dead, betray on the part of this little known people a love of nature and an appreciation of its beautiful features which are to be classed among man's nobler faculties, and which cannot fail to excite in us some tender sentiments mingled with our curiosity to learn something of their lost history. The grave mounds or barrows of Great Britain

were located with a similar regard to beauty of outlook and conspicu-
ousness. Llewellyn Jewett, in "Grave Mounds and their Contents,"
writes as follows :

"The situations chosen by the early inhabitants for the burial of their
dead were, in many instances, grand in the extreme. Formed in the
tops of the highest hills, or on lower, but equally imposing positions, the
grave mounds commanded a glorious prospect of hill and dale, wood and
water, rock and meadow, of many miles in extent, and on every side,
stretching as far as the eye could reach, while they themselves could
be seen from afar off in every direction by the tribes who had raised
them, while engaged either in hunting or other pursuits. They became,
indeed, landmarks for the tribes, and were, there can be little doubt,
used by them as places of assembling."

Permission having been generously granted by the owner of the
property, Mr. Chas. Roberts, to make such use of the mound as the
Association should see fit, it was at first proposed to remove it layer by
layer with plow and scraper, in order to expose its whole floor at once,
but a careful survey soon made it apparent that such a mode of pro-
cedure would be out of the question owing to the number of trees,
some of considerable size, scattered over its surface, and their inter-
lacing roots. Work was accordingly begun by carrying an adit from
the northwest side and sinking a central shaft of the dimensions of
about four by eight feet, the longer diameter of the shaft running
north and south. In the side adit, which was dug first, nothing was
disclosed till the floor of the mound was reached, when just before
coming to the natural surface of the soil, perhaps a foot above it, the
trench passed through a layer of white ashes. This layer was afterward found to extend from nearly the outer margin of the base of the
mound, across its whole floor, arching up over the center so as to present a convex surface above. Its thickness varied from half an inch
to one and a half inches. The layer presented near the center of the
mound an almost stony hardness, causing it to come off in large flakes
to which masses of the surrounding clay often adhered. When the
clay was cleaned off the layer disclosed a mottled surface of a reddish
brown color. The hardness was apparently due to the lime of which
the ashes seemed to be mainly composed, and the reddish brown surface
might have been produced by a covering of bark placed over the ash
layer, or more probably in the manner related in the following passage
from "Grave Mounds and their Contents," describing the mode of
burial probably practiced by the ancient Celts:

"It not unfrequently happens that the spot where the funeral fire has
been lit can very clearly be perceived. In these instances the ground

beneath is generally found to be burnt to some considerable depth; sometimes, indeed, it is burned to a fine red color, and approaches somewhat to brick. When it was intended that the remains should be collected together, and placed in an urn for interment, I apprehend from careful examination, that the urn, being formed of clay,   *   *   * was placed in the funeral fire, and then baked, while the body of the deceased was being consumed. The remains of the calcined bones, the flints, etc., were then gathered up together and placed in the urn ; over which the mound was next raised. When it was not intended to use an urn, then the remains were collected together, piled up in a small heap, or occasionally enclosed in a skin or cloth, and covered to some little thickness with earth, and occasionally with small stones. Another fire was then lit on the top of this small mound, which had the effect of baking the earth, and enclosing the remains of calcined bones, etc., in a kind of crust, resembling in color and hardness a partly baked brick. Over this, as usual, the mound was afterward raised." Page 83.

A chemical analysis will doubtless throw farther light upon these points. Below this layer at a varying depth, but of 8 or 9 inches on the average, a second layer was reached similar in character to the upper one. The space between was filled with clay like that composing the mound. The relative portion of these two layers is seen in the accompanying diagram (Plate 1, Fig. 6). At the point of junction between the side adit and central shaft, was found a heap of loose ashes mingled with small fragments of calcined human bones, (Fig. 8, D). In the heap were found also, several rudely fashioned flint arrow heads and a pierced ornament of stone. These seemed to have been acted upon by heat, as if some warrior, with his ornaments and weapons upon him, had been incinerated and the remains carefully collected and deposited where found. At a later day one or two other heaps of calcined bones were found, all at about the same distance from the center of the mound, (Fig. 8, E). The point to be determined is whether the whole of the ash layers was not originally composed of burnt bones. In carrying down the central shaft some fragments of human bones, much decayed, were unearthed near the surface, marking the site of an *intrusive burial*. Some scattered fragments of calcined bones were also found which will be referred to farther on. At the depth of from three and a half to four feet near the center of the mound a human skeleton was reached, lying on the back, the head toward the north, (Fig. 7). It was firmly imbedded in the compact clay, and so great was the care required in removing it that only the head and upper part of the trunk was secured before night came. It will be proper to mention that during this first day's work, the writer was assisted by Mr. John B. Niles, of Urbana University, who also gave valuable aid during the whole exploration. The following Monday the work commenced on Saturday was resumed. The party this time was increased by the addition of Mr. Aaron Aten, of the Association, and Messrs. Roberts, Bacon and Cabell, Students at the College. We were also favored with the presence of Mr. Galen C. Moses, of Bath, Maine, a gentleman much interested in archæological matters, who showed his zeal by coming more than a thousand miles, expressly to be present at the opening of this mound. He remained with us two days and contributed much to the success of the work by his practical suggestions. The first operation was the uncovering of a space some twenty feet long by eight wide, and to the depth of three and a half feet, in which nothing was exposed save a quantity of burnt clay and charcoal, to which reference will again be made. The work of disinterring the remainder of the skeleton discovered on Saturday, was soon resumed and after an hour and a half of hard labor, conducted with great caution, we had the satisfaction of securing almost the entire skeleton, only a few bones of the ankle and wrist and several phalanges being wanting. It should have been mentioned above that in making the preliminary excavation, a handsome and perfectly formed perforated stone ornament was found, which, however, seemed to have no connection with the burial, as it lay at a considerable distance from the skeleton and not on the same level with it. This skeleton has been found to weigh exactly nine pounds.

A more particular description of the crania and bones found is reserved for a subsequent paper. It is an interesting fact that the breast bone has been perforated by some sharp instrument, probably a flint spear or arrow head, as the aperture is much larger on the outer than upon the inner surface of the bone, showing it to have been made by a tapering instrument. The external opening measures $1\frac{1}{2}$ inches while the internal one is but $\frac{3}{4}$ of an inch. Under the right thigh, near its upper third, was found a fragment of quartz rock as large as the palm of the hand, one side of which was flat and polished, the polished surface extending in the arc of a circle over one of the edges. Unfortunately this relic was misplaced, and could not afterward be found. This concluded our "finds" for the day. The next day, Tuesday, two workmen were set at work in the morning to deepen the excavation, under the superintendence of Mr. J. B. Niles, the writer, on account of College duties, not being able to reach the scene of operations until noon. In the afternoon the skull of skeleton No. 2 was reached, (Plate 1, Fig. 8, B). This skeleton was in a very imperfect condition, and but a small part of it was removed. As far as could be judged from the position of the scattered fragments it had been placed upon the back, with head toward the west. It was underneath the upper layer of ashes, and the head was but a short distance from the heap of calcined bones first described. At a short distance from it the bones of another skeleton .were found which we will designate as No. 3, (Fig. 8, G), as parts of the one were unearthed simultaneously with parts of the other and all presented a very confused arrangement. The fact of there being but two skeletons was not ascertained at the time, but only disclosed after several hours work, at a later day, in arranging and comparing the fragments. The position of the parts of No. 3 was indeed such as to seem to render it conclusive that the bones had been gathered together and burnt side by side, but after study renders it not impossible that the body had been folded together and laid on its side. The bones of the forearm and hands are, however, entirely wanting, and but very little of the spine is present. This might have been the result of decay but the bones that remain are remarkably heavy and nearly vitrified. They were covered over with a thick incrustation and presented the appearance of having been in the fire. This crust could not be removed at the time, but I find that after exposure to the air for some days, it has a tendency to scale off. One of the small bones of a leg and the upper bone of an arm lie across each other and are firmly-attached. The articular end of the long bones are nearly all absent in this skeleton, the lower jaw is much awry and the skull has a very low and retreating forehead. Altogether it presents the appearance of a very low type of humanity. Near by a gouge or spoon, hollowed out of the metatarsal (?) bone of the elk (?) was found This was also incrusted in a way similar to the bones above described. Implements of bone are frequently found among mound relics and they constitute an important group in the classification of prehistoric articles. I have now mentioned nearly all that has been found in this mound up to the present time. (Plate 5, Fig. 5).

Some two feet from the surface, at the south end of the excavation, a mass of charcoal was met with the fragments of which were of large size. With the charcoal was found a piece of thigh bone, charred and petrified, and part of an ulna or bone of the forearm. Near by was a stratum of clay, burnt nearly red; at the side, however, and not resting upon the charcoal. It is inferred that these charred fragments of bones and charcoal, as well as those found on the surface, were scraped up from the site of the cremation, having been left behind when the ashes were gathered for burial, and that they were thrown on the mound with the surrounding earth during the process of its construction.

In this mound two modes of burial appear to have been simultaneously practiced, viz: those of inhumation and cremation, unless we are to regard the imperfect skeletons on the floor of the mound as belonging to the latter, the operation being but imperfectly performed. The condition of some of the bones hardly justifies such a conclusion. Mounds containing examples, both of inhumation and cremation, were of frequent occurrence in Great Britain, and many of them are described in Mr. Jewett's "Grave Mounds and their Contents." The skeleton found nearer the surface was of course deposited at a later date than the others, and may have been an intrusive burial, or it may have been deposited upon the former surface of the mound and the mound afterwards increased in size. The character of the bones themselves must be our only guide in determining this point. The practice of cremation, sufficiently common in ancient times, is still observed to some extent by the native races of North America, though entirely foreign to the spirit pervading the funeral rites of modern white races, as is shown by the attention excited by the two or three recent instances of it. A paper, read before the meeting of the American Association for the Advancement of Science, will serve to illustrate this mode. (See p. 58, "American Naturalist," for 1875). Some tablets recently found in a mound in Iowa, by the Rev. J. Gass, give a pictorial illustration of the process. (See p. 109, vol. 2, part 1, Proc. Davenport Academy of Sciences.)

THE BALDWIN MOUND.

This mound is located upon the top of a hill lying between the north and east forks of Buck creek at their junction, some eight miles southeast of Urbana and upon the farm of the late Judge Samuel Baldwin. It is nearly conical in shape, about seventy-eight feet in diameter at the base, and fifteen in height. Upon it oak trees of considerable size are now standing. The south side of the mound shows evidence that a considerable portion had at some time been removed. I am informed by Mrs. Baldwin that this appearance is due to the fact that the clay from the mound was used to make the brick for the house now standing upon the farm and occupied by Mr. Frank Baldwin. This house, a two story brick, was built some fifty years ago. In the process of removing the clay it is said that a quantity of bones was unearthed but afterward re-interred. Work was begun by carrying an adit from the side towards the center, and after the center was reached sinking a shaft toward the base. Some two feet from the surface the bones of several skeletons were found. These are frequently found in the surface of mounds, and are generally accounted to be those of some Indian tribe and of comparatively recent date. At the depth of twelve feet the original place of sepulture was reached. Here a rude structure of bark and branches had been made as a receptacle for the dead, constructed, as nearly as could be ascertained, in the following manner: First a layer of bark was laid down, then the bodies placed upon this, the head of the one being directed toward the east, of the next toward the west, and so on. Logs were placed at the sides and between the bodies, dividing the grave into as many compartments as there were persons to be buried. The whole was then covered with a thick layer of bark, upon the surface of which was found a thin layer of charcoal. Bark, branches and bodies had of course reached the last stage of decay, only the ashes of the former remaining to show how they had been disposed, and long hollow cavities filled with dust alone indicated the position of the logs. The whole mass had been pressed down and flattened by the weight of the overlaying earth, and most of the bones showed evidences of the great pressure, being crushed in and broken. The first skeleton reached was found lying with head toward the east, and it was judged to be that of a female. In the vicinity of the pelvis a number of bones of the head and limbs were found evidently belonging to an unborn child, judging from the condition of the teeth in the lower jaw, and that of other parts of the skeleton. Some of these bones are figured of their natural size in Plate 1, Figs. 3 and 4, and also in Plate 3, Figs. 3 and 4. (In Plate 1 Fig. 3 represents one of the bones of the toes, and Fig. 4. several of the fingers. In Plate 3 the bones of the thigh and leg are seen). A small copper ring was found near the head which had probably been worn in the ear or nose. (Plate 2, Fig. 2). Further excavation disclosed a second skeleton, having its head directed to the west. The bones of this skeleton were evidently those of a warrior, being very large and strong, and those of the lower limbs were in a remarkable state of preservation. Near the hand and lying across the body were the flint heads of three spears or arrows.

Their position seemed to show that they had been held in the hand by the wooden shafts now mouldered away. The upper part of the body when exposed to view presented a remarkable appearance, being crushed and distorted to a great extent by the pressure above. It had apparently been placed upon its left side and the arrows were grasped in the right hand. Removing the earth carefully from this a third skeleton was seen, its head pointed to the east. This was lying upon its back, and measured from its toes to the top of the head nearly six feet; but this measurement cannot be considered perfectly reliable, owing to the flattening of the body from pressure. The teeth were thirty-two in number and perfectly sound. Around the neck was a string of beads made of mother of pearl, probably taken from the shells of the river mussel. (Plate 2, Fig. 3). This skeleton seemed to be that of a young woman of from 18 to 20 years. (Plate 4, Fig. 1 represents the profile and Fig. 2 the face view of the skull).

The skeleton next disclosed was that of a young man of about sixteen years. The head was placed in the reverse position to that of the one preceding it. The skull of this one is remarkably well shaped as may be seen by reference to the illustration, (Plate 3, Figs. 1 and 2). Over the breast were found several plates of mica, cut in the form of a crescent, (Plate 2, Fig. 1). Plates of mica, often of large size, are frequently found in mounds and the mica is believed to have been brought from Carolina. This, with the copper from Lake Superior and the small shells from the Gulf of Mexico about to be mentioned, is an evidence of the commercial habits of the people. The next space was occupied by the skeletons of two small children placed feet to feet. Near the head of one of these was a heap of small sea shells, belonging to a species now found in the Gulf of Mexico. These were pierced at the ends and may have been worn as a necklace, (Plate 2, Fig. 9). The succeeding skeleton was that of an adult person and near it was found a small implement made of banded slate, belonging to the class called "boat shaped" implements in the collections of the Smithsonian Institution. It may have been worn as a mark of dignity or badge of office. It is not pierced by holes, (Plate 2, Fig. 5). An eighth skeleton was found belonging to this group, near which also lay a small quantity of shell beads like those last described. Following these, near the margin of the mound, were three others thrown down apparently without regard to position, as they were disposed at various angles with the limbs crossing each other,

and no protection of logs had been placed arround them, nor were any
ornaments found with them. It is worthy of note that of all the
skeletons found in the mound the eight first described were buried
with especial care and each of them had some mark of distinction or
token of affection. The arrangement of the bodies was also somewhat
remarkable, they being placed with great uniformity with the heads
alternately toward the east and west. The suggestion is offered that
this arrangement was made simply with a view to economise space.
During the last war the writer had occasion to visit the battle field of
Chantilly, Va., for the purpose of obtaining the body of a relative
killed in that battle. There had been a hasty retreat and but little
time was afforded for the burial of the dead. The body sought was
found in a shallow pit evidently made in haste, and in this pit 37
bodies had been placed and barely covered with earth. The arrangement of the bodies was precisely the same as in this mound, heads and
feet being placed alternately, with the evident object of crowding as
many as possible into the excavation. This disposition of the bodies
allows a greater number to be placed side by side than if all had been
deposited with the heads in one direction.

The excavation above described occupied the northwest quarter of
the mound and nothing farther was found in this section. Some days
later work was resumed by carefully uncovering the whole of the
northeast quarter·of the mound until the floor was reached. The
space was found covered with skeletons, 10 in number, which had
been promiscuously thrown down, the bodies being bent at all angles
and the limbs of one often lying across those of another. As in the
case of the former deposit a layer of charcoal was found over the upper
surface and another had been placed below. The cross section of the
mound plainly exhibited these two layers of charcoal. No implements
or ornaments were found with these bodies. In one corner of the
area, near the center of the mound (Plate 1, Fig. 5, X.), there was a
small heap of ashes containing a few burnt bones and calcined mussel
shells. At the outer angle (Fig. 5, V.), was a vase of baked clay
crushed to fragments, the rim only retaining its original form. This
vase or funeral urn was placed with the mouth downwards. Its interior surface was coated with black carbonaceous matter. In the
south half of the mound, that part from which the clay had been
taken for making brick, a pit was dug and bones were reached quite
near the present surface. Here were parts of three skeletons, and as

no others were found near by, it was conjectured that these might be the remains of those which had been encountered at the time of making the brick, and which were said to have been again buried. The excavation was continued toward the center of the mound without farther results, but this may have been from not carrying it to a sufficient depth. A test pit was also made on the west side, but nothing was found. The plan of the several excavations and of the deposits found is shown in Plate 1, Fig. 5. The vase is represented in Fig. 1. The shape there given is somewhat conjectural, as the upper part could not by any possibility be restored, owing to the small size of the fragments and the crumbling away of their edges. That the general form was rounded is known from the shape of the fragments. Fig. 5, Plate 3, represents a fragment of what appears to be the metatarsal bone of some species of the deer family. Comparison has been made with the same bone in the elk and other species of the deer family, but a remarkable variation is observed in the proportion of the parts. In this specimen the width of the outer condyle is much greater in proportion to its thickness than in the case of the recent specimens. An implement of bone having a sharp cutting edge and polished sur- was also found near the first group of skeletons. This is shown in Plate 2, Fig. 4. Figure 9 of the same plate represents a small flint instrument of unusual form. It is represented of natural size. Figures 6, 7 and 8 are of objects not found in this mound and are described elsewhere. It is quite probable that much more remains to be discovered in the remaining portions of the mound, and it is the design of the Association to continue the exploration at some future day. The soil of which the mound is composed is a clear yellow clay, quite free from stone or gravel, and cutting under the spade with a smooth bright surface. The hill upon which it rests is a loose mixture of limestone pebbles having a thin surface covering of dark loam. For a considerable distance around the base of the mound the earth is somewhat similar in character to that of the mound, but is not so free from stone. Such pebbles as were found in the mound were of quartz and sandstone, only a single specimen of limestone being found and that a water worn one. The material did not appear to have been scraped up from the surface, but was probably brought from a distance. The clay was nearly homogeneous throughout and very compact. At the base a complete arch had been formed by the decay of the log structure, the superincumbent soil having first become sufficiently firm to retain its position. This peculiar structure of the mound was the cause of what was well nigh a serious accident. While stooping down near the center of the main excavation to examine a heap of ashes and being at the time some ten feet below the surface, one of the excavators, the writer of this article, was suddenly buried by the caving in of the side wall. The blow was a terrible one and for a few moments the situation seemed critical, as respiration was nearly impossible, owing to the weight of the mass of earth that had fallen. Through the exertions of Dr. J. L. McLain, Mr. Frank Showers and Mr. Corbin, who were assisting in the excavation, a timely rescue was effected with no serious result. The incident is mentioned in the hope that it may be of service to those engaged in similar work, and as a wholesome caution, especially as during the same month a similar accident occurred in the process of opening a mound in West Virginia, which had a fatal result.

It was the intention to append to this account of the opening of the mound a description of the crania and other bones, of which a large number was secured, but these will be made the subject of a separate paper. In general it may be said that the same peculiarities that are mentioned by other writers, such as the foreshortening of the skulls and their want of symmetry, the flattening of the tibia and perforation of the humerus are all exhibited in a marked degree. Many of the bones had become bent by the weight of earth resting upon them and much of the distortion exhibited in the skulls is believed to be due to this cause rather than to compression during life. This was proved in some instances in attempting to replace the bones of skulls that had fallen apart, a wide gap remaining at the completion of the work between parts that should have come into contact and which the most skillful manipulation could not conquer. To account for this bending and warping of the bones it is only necessary to consider the peculiar conditions to which they have been subjected. For centuries they have lain beneath an immense mass of earth, and this constant and long continued pressure, accompanied by a kind of molecular disintegration and re-arrangement of the particles of bony matter, is amply sufficient to produce these changes of form. Frequently bones lying over each other were found soldered together from the same cause.

OTHER MOUNDS.

A number of mounds, of less important character then those just described, have been opened from time to time by the Association.

A small flat mound at the junction of the Mackachack with Mad
river on Mr. Clem's farm disclosed nothing but a layer of charcoal
near its base. If it had contained any bones or other objects of interest, these had either decayed or were not reached by the line of excavation. A mound of similar character was also opened on the farm
of Mr. Michael near Buck creek. In this nothing but a small quantity of charcoal was found.

Two mounds have been opened by the Morse Natural History
Society in which work the writer was also associated. One of these
is located on the ridge northeast of the Baldwin mound on the farm
of Mr. Wilson. It is a low mound, being only three feet in height
and about thirty in diameter. In it were found simply the frontal
bone of a skeleton, fragments of a stone pipe and a few flints. A
mound has been recently opened by the same society with the assistance of James Bacon, Esq., of Springfield, which has yielded interesting results as far as the exploration has been carried. This mound is
situated on the Foley farm, about four miles east of Springfield on a
ridge midway between Buck creek and Beaver creek. The mound is
eight feet in height and seventy feet in diameter at the base. It is
situated in a wood and is covered by the present forest growth. An
excavation eight feet square was carried down through the center to
the base. The line of junction of the mound with the ground on
which it stands is very well marked. Four feet from the top, a mass
of burnt clay was encountered in which were found fragments of
charred bone and the skull and other parts of a skeleton that had not
been charred. Bits of pottery were met with here and there during
the whole of the excavation. The principal deposit was reached at
the base of the mound immediately upon the surface of the ground.
Here were several skeletons variously disposed. They were placed in
a small area close together and apparently without any enclosure of
wood, bark or stone. There was no charcoal about them, and the
bones were firmly imbedded in the clay. The first skeleton exhumed
was folded together, the knees resting against the chin, and was placed
on the right side, the head toward the north. Next to it was a skeleton in a similar position with the head toward the south. Beside
these, stretched at full length, were two others. The first was placed
on its back with the head turned on its left side toward the north and
facing the east. Close behind this was another skeleton in a similar
position to the former, the skulls of the two being in contact and the

right arm of the latter lying across the breast of the former. Near the pelvis of the last one were found a quantity of very small human bones, apparently those of a fœtus. The three bones of the ossa innominata were found separated, never having become united. Near the feet of the two skeletons was found a few handfuls of ashes mingled with fragments of mussel shell. This last deposit was similar to one found in the Baldwin mound and probably had something to do with the funeral rites. The skulls of two large rodent animals were also found near by. The skull and most of the long bones of the last or female skeleton were saved entire. The skull is similar in shape to the others which have been exhumed, having the same cuboidal form and vertical occiput which are characteristic of the genuine mound skeletons. Up to the present time no art relics have been discovered save the fragments of pottery mentioned above. Several other mounds are located in this neighborhood on the banks of Beaver creek.

Several miles below Springfield at Snyder's Station on the A. & G. W. railroad, east bank of Mad river, a mound was opened by John Snyder in which was found a cache containing 128 flints of the "leafshape" pattern of various sizes. The flints are similar to the ones in the possession of Mr. Aaron Aten and figured in Plate 5, Figs 4 and 5. Mr. Aten's were plowed up in a field in Wyandot county and are 182 in number.

A mound was recently opened upon the farm of the late Judge Dallas, four miles below Urbana, by Mr. James Dallas, who has exhibited to me the results of his exploration. This mound is placed on the summit of the bend overlooking Mad river valley, and as the valley here changes its direction, making a sweep towards the southeast, and is some three or four miles in width, the situation is a very commanding one. The beauty and extent of the view from this mound is remarkable. The mound, from a survey made by Dr. R. H. Boal, was found to be fifty feet in diameter and four feet in height, and situated 105 feet from the edge of the plateau. A few rods below, on the slope of the hill, is a small circular ridge, some fifteen feet in diameter, the earth forming the ridge being thrown out in such a way as to leave a small conical elevation in the center. The relics obtained from the mound and now in Mr. Dallas' possession consist of several boat shaped badges, a number of arrow heads and leaf shaped flints, a fragment of a stone tube and in addition to these articles a quantity of charred bones, among which was a mass of charred cloth. This last interesting deposit will be described in detail elsewhere. One of the badges is much longer and more tapering than the others and is the only one pierced with holes. It is made of banded blue stone, and as new and fresh as if it had just been wrought. It is a matter of regret that the opening of the mound had not been made by some experienced person so that the relative position of the cloth and bones might have been observed.

The places of sepulture over which no structures have been erected are exceedingly numerous throughout the region we have been describing. Hardly a railroad or turnpike cutting is made, or gravel bed opened, that does not disclose a mass of skeletons. The hill tops are literally sown with the dead. It is probable that these burials belong to all periods in the history of the country down to the time of its occupancy by Europeans. The mound builders may have buried their common dead in this way, reserving tumuli for their chiefs and their families and for other distinguished persons.* The two modes of burial are, it must be observed, entirely different from each other. In the case of the mounds the remains are placed upon the surface of the ground, often being first incinerated, and the earth heaped over them, while in the latter instances, a pit is dug and the remains are interred below the surface. In the case of these gravel bank burials every surface indication of the cemetery below has usually been effaced by time. The bones are found in the knolls, quite near the top. Generally, when exposed, the depth of the deposit may be known by the mixture of the surface loam with the gravel under-neath. The bodies may occur single or in graves grouped together, or crowded promiscuously into large trenches. They have been found in almost every posture, prostrate, sitting and even standing.† Sometimes the parts of the same skeleton are quite widely separated from each other and so mingled with the materials of the drift that they would almost seem to have been deposited by some surface action before the alluvium was laid down upon it. This condition of things was especially observable in a deposit examined by Dr. J. L. McLain, Prof. J. E. Werren and myself, on the farm of S. M. Hodge on the east side of Buck creek valley and near the southeast corner of Champaign county. The bones of portions of several skeletons were found

*Ancient Remains of the Mississippi Valley p. 171.

†An instance of this last posture was related to me by an eye witness, but I do not know how reliable the statement may be.

on the side of a gravel knoll, four feet from the surface, completely imbedded in the gravel, which showed no trace of the superficial soil. In the case of one of the skeletons the lower jaw was removed two feet from the skull to which it belonged, and the bones generally were much scattered. It is an interesting fact that a thigh bone, having a perfectly re-united fracture, was found among the bones. The position of the united parts is bad, and the amount of shortening is upwards of three inches. A few rods distant on the summit of the knoll stands a small mound; there is also another in the bottom and still another on a hill on the opposite side of the stream. Many flints are plowed up in the neighborhood.

A large deposit of skeletons has been unearthed at Catawba station, on the C. C. C. & I. railroad, in the work of removing a portion of the steep gravel bank at this place for ballasting the road bed. Rev. G. G. Harriman reports that the bones were found buried in two trenches, running east and west, eight feet apart and eighteen inches below the surface, resting on the top of the gravel. There was no appearance of any regular burial, but the bodies had apparently been thrown in helter skelter. They comprised both those of children and adults. Most of the bones fell down the bank during the work of excavation and were carried off on the gravel trains. The place was visited by parties from Delaware, Springfield and Cincinnati, who all secured specimens, and a skull and several of the long bones in a good state of preservation were obtained for the cabinet of the Association. I examined on the spot as many of the bones as I could gather and observed that a large percentage of the humeri, or arm bones, was perforated. The skulls seen are small and rounded in form, much resembling those found in the mounds of the neighborhood. It was estimated that several hundred skeletons had been exhumed from these trenches. The hill on which these remains were found is the eastern border of a jutting portion of the plateau or bench which borders Buck creek valley. This extends out into the valley on the west like a kind of promontory, and on the end of the promontory are three circular depressions. From these depressions a graded way, as wide as an ordinary wagon road, leads to the foot of the hill. The attention of the Association was called to these depressions by Mr. Joseph C. Glenn, upon whose farm they are found.

An account of the antiquities of Mad river valley would not be complete without a map which should contain the location of every mound and earthwork at present existing in this region. The construction of such a map is contemplated when the necessary materials for it have been collected. Finally, it is not out of place to mention here the existence of another class of ancient remains, namely, those of the mastodon, which are quite frequently found. Within a few years not less than three deposits of this character have been discovered in the peat bogs bordering the river. These were casually discovered in the work of ditching and doubtless the remains of many more are scattered through the valley.

EXPLANATION OF PLATES.

48

PLATE III.

Fig. 1. Skull from Baldwin Mound. Profile.
Fig. 2. Same, face view.
Fig. 3. Thigh bone of infant, Baldwin Mound. Size of
nature.
Fig. 4. Bones of the leg, same as above.
Fig. 5. End of metatarsal bone of elk, found in Baldwin
Mound. - - - - - - - - Page 37.

PLATE IV.

Fig. 1. 2. Profile and face of skull from Baldwin Mound.
Fig. 3. Small bones of the ear, Baldwin Mound. - - Page 37.

PLATE V.

(All articles in this Plate reduced one-half).

Fig. 1. Badge of banded blue stone found on the bank of
the Broken Sword, in Eden Township, Wyandot
Co., O. Collection of A. Aten.
Fig. 2. Badge, or "Ceremonial Ax," from Antrim Tp., Wy-
andot Co., O. Collection of A. Aten.
Fig. 3, 4. Leaf shaped flints being the largest and smallest
of a lot of 182 plowed up in a field in Antrim Tp.,
Wyandot Co., were placed in a pile, the largest at
the top. Collection of A. Aten.
Fig. 5. Implement of bone from Roberts Mound. - - Page 37.
Fig. 6. Plummet or Sinker. Collection of A. Aten.

PLATE VI.

Fig. 1. Stone Pipe from Champaign Co. T. N. Glover.
Fig. 2. Stone Pipe. Cabinet of Urbana University.
Fig. 3. Cast of Clay Pipe found near Woodstock. T. N.
Glover.
Fig. 4. Sandstone Pipe resembling Toucan, found in Cham-
paign Co. Property of C. Johnson.
Fig. 5. Sandstone Pipe from Pipe Town, plowed up in
1871. Collection of A. Aten.
Fig. 6. String of ivory beads or wampum. From Mrs.
Major Hunt.

PLATE VII

Fig. 1. Double Crescent, or "ceremonial weapon" of banded
slate. From Antrim Tp., 3 miles below Upper
Sandusky, O. (Reduced ½). Collection of A.
Aten.
Fig. 2. Stone Ax from bank of the Broken Sword. (Re-
duced ½). Collection of A. Aten.

PLATE 1.

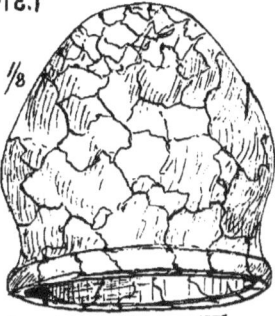

FIG.1

1/8

Baldwin Mound Vase.
(Restored)

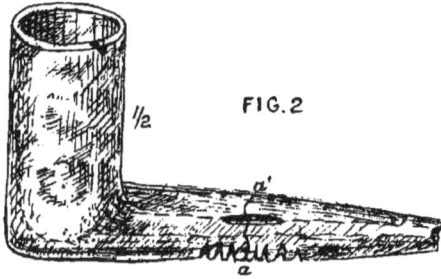

FIG.2

1/2

FIG.4

FIG.3

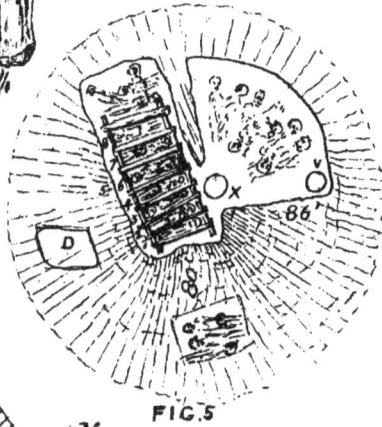

FIG.5

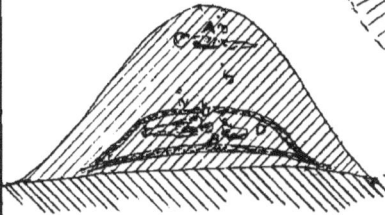

FIG.6

N.

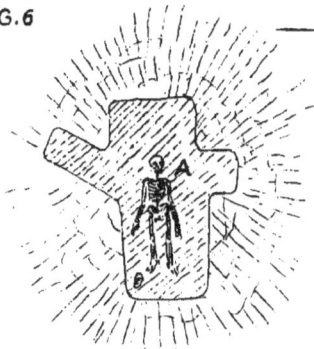

FIG.7

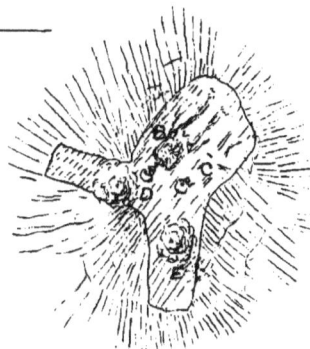

FIG.8

D.H.Sherman Del.

THE HELIOTYPE PRINTING CO. 220 DEVONSHIRE ST. BOSTON

PLATE 2

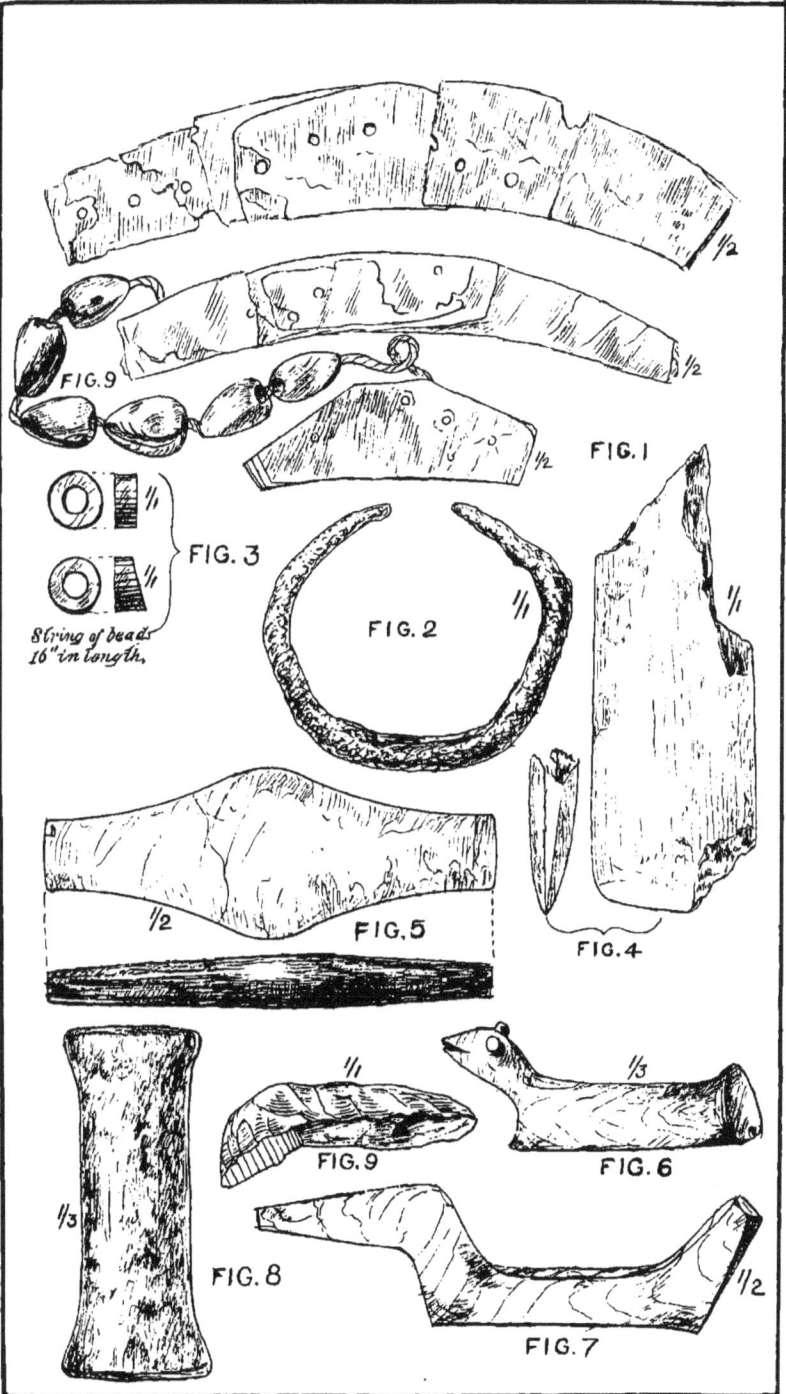

FIG. 9

FIG. 1

FIG. 3

String of beads
16" in length.

FIG. 2

FIG. 4

FIG. 5

FIG. 9

FIG. 6

FIG. 8

FIG. 7

D. H. Sherman Del.

THE HELIOTYPE PRINTING CO. 220 DEVONSHIRE ST. BOSTON

PLATE 3

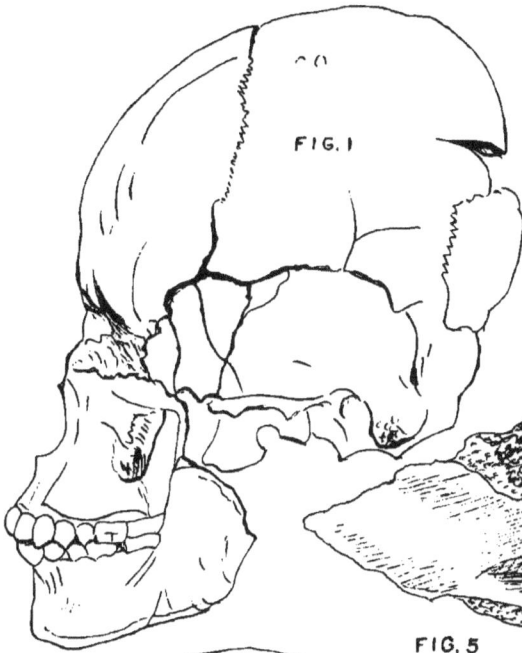

FIG.1

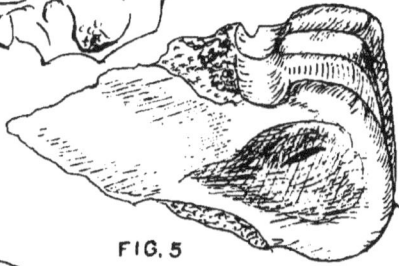

FIG.5

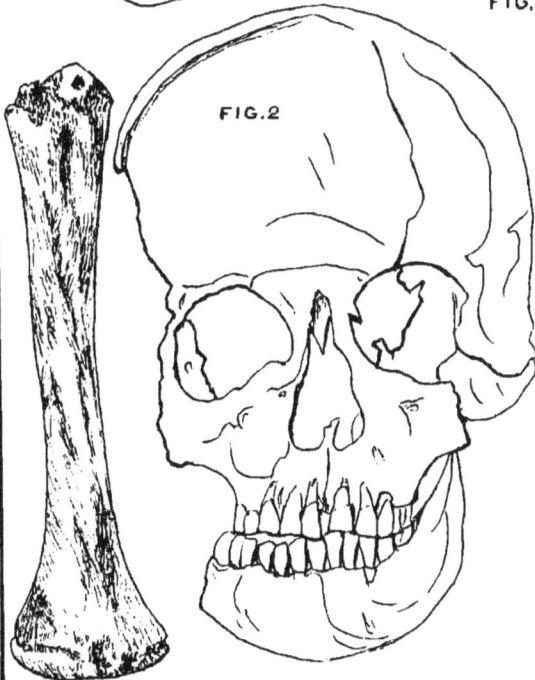

FIG.2

FIG. 3

FIG.4

D. H. Sherman Del.   THE HELIOTYPE PRINTING CO. 220 DEVONSHIRE ST BOSTON

PLATE 4

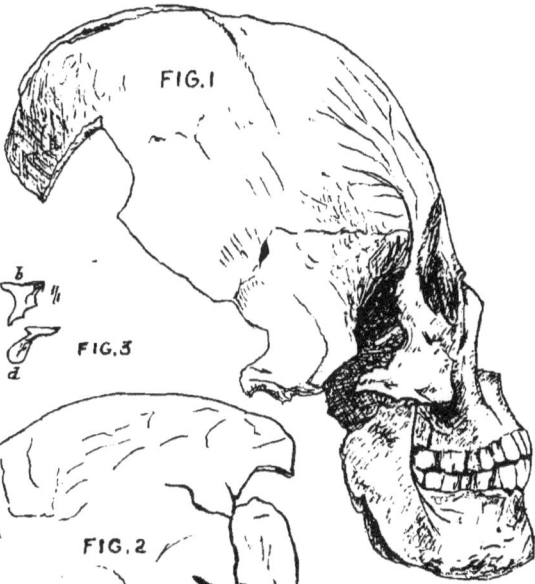

FIG.1

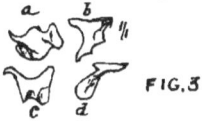

FIG.3

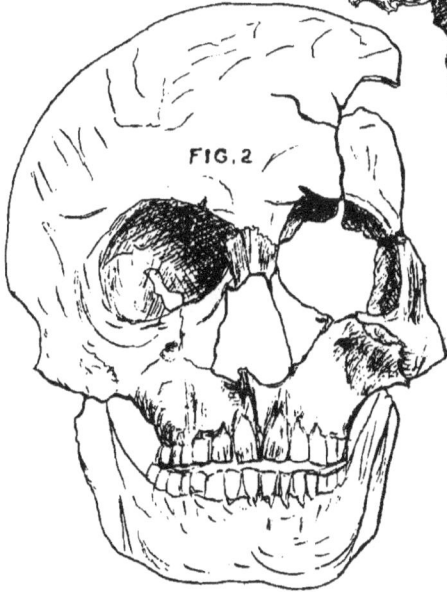

FIG.2

THE HELIOTYPE PRINTING CO. 220 DEVONSHIRE ST. BOSTON

PLATE 5

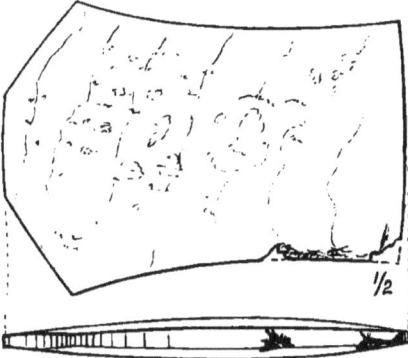

FIG. 1

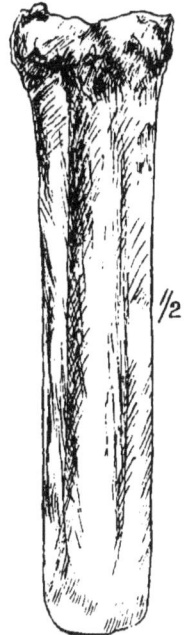

FIG. 5

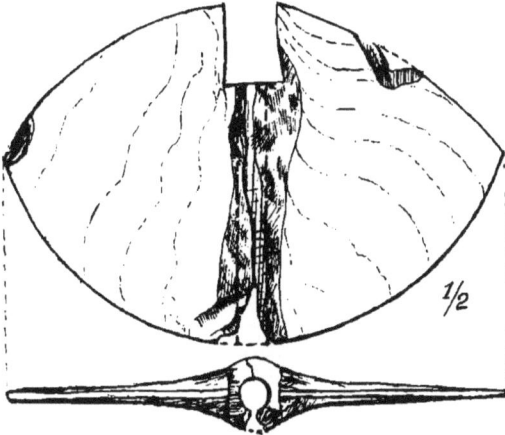

FIG. 2

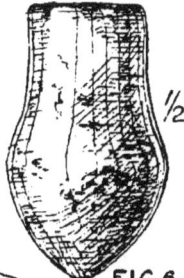

FIG. 6

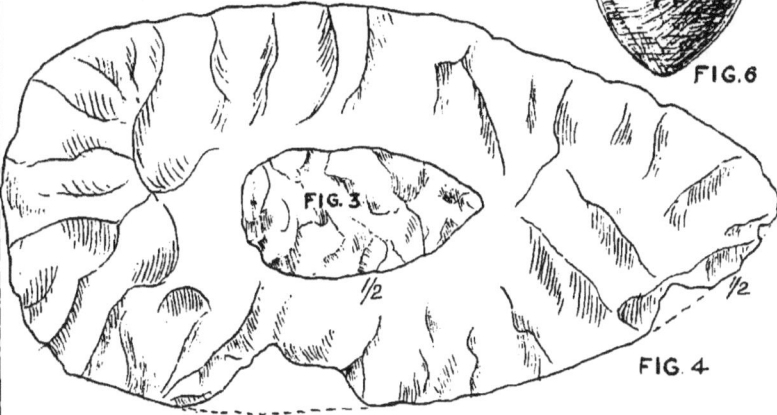

FIG. 3

FIG. 4

PLATE 6

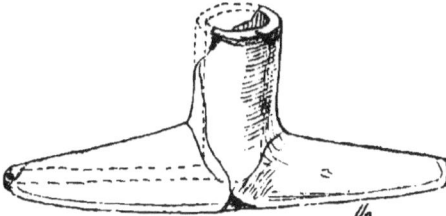

FIG.1   ½

FIG.2

FIG. 3   ½

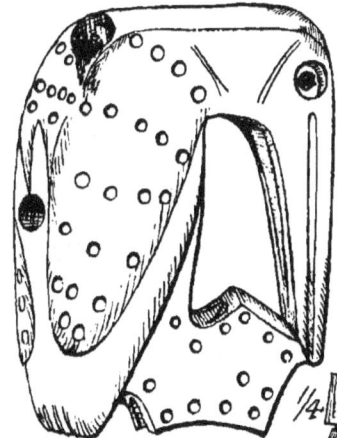

¼

FIG. 4

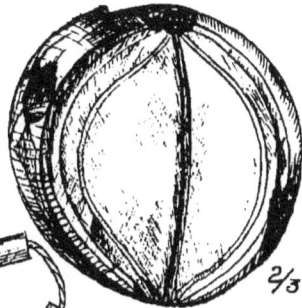

⅔

FIG. 5

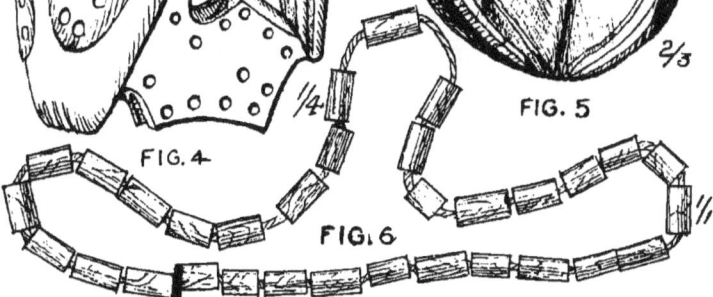

FIG. 6   ⅟₁

D.R.Sherman Del.   THE HELIOTYPE PRINTING CO. 220 DEVONSHIRE ST. BOSTON

PLATE 7

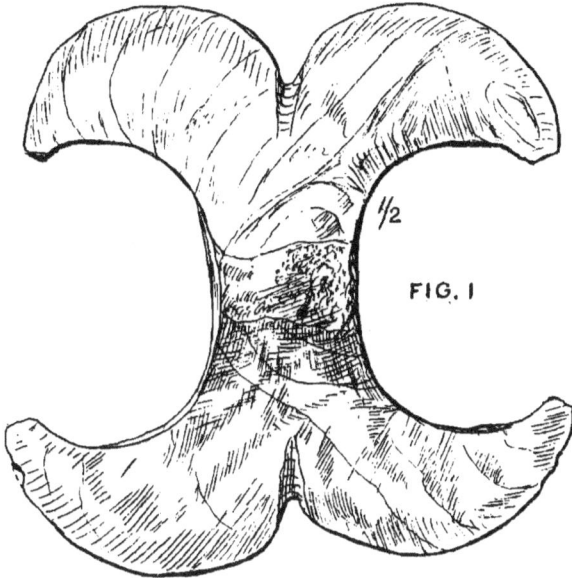

½

FIG. 1

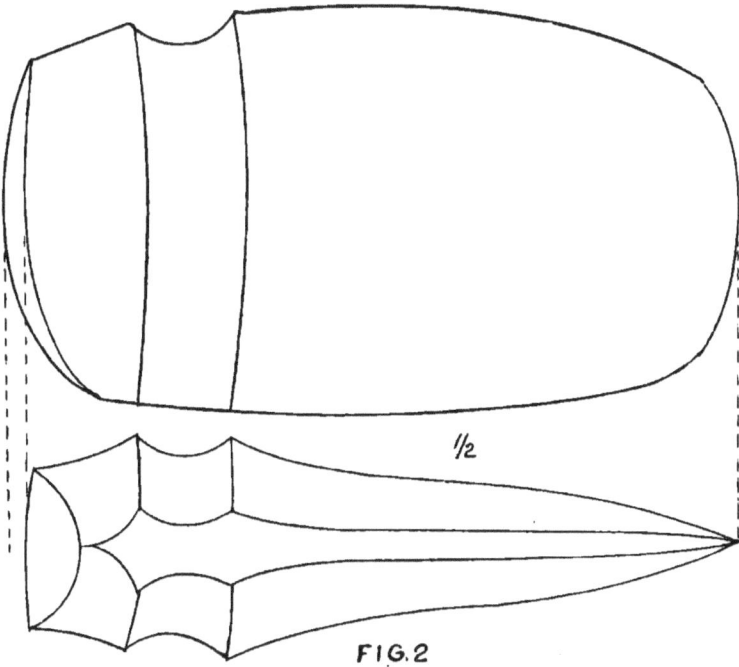

½

FIG. 2

D. H. Sherman Del.

THE HELIOTYPE PRINTING Co. 220 DEVONSHIRE ST. BOSTON.

PLATE 8

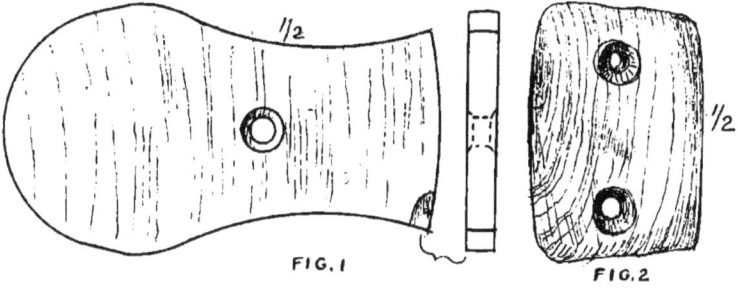

FIG. 1

FIG. 2

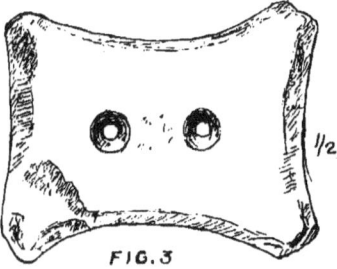

FIG. 3

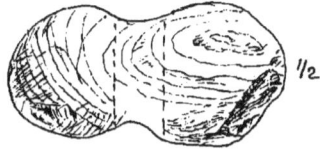

FIG. 4

FIG. 5

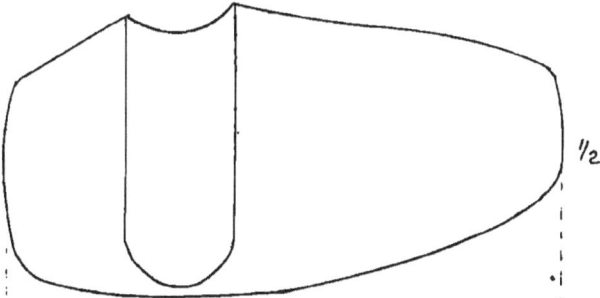

FIG. 6

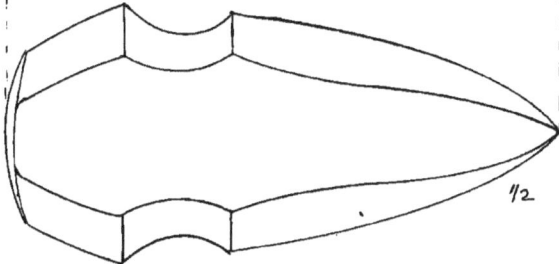

D. H. Sherman Del.

THE HELIOTYPE PRINTING CO. 220 DEVONSHIRE ST. BOSTON.

49

PLATE VIII.

(Figures reduced one-half).

Fig. 1, 2, 3. Stone badges or "pierced tablets." Cabinet of Urbana University.

Fig. 4. Stone badge. Cabinet of Urbana University.

Fig. 5. Badge of banded slate from Antrim Township, Wyandot Co. Collection of A. Aten.

Fig. 6. Perfectly formed ax of fine grained syenite, found in Clarke Co. Collection of T. F. Moses.

ALTITUDE OF THE BLUE LIME-STONE FORMATION AT OSBORN, O.

BY S. F. WOODARD, OSBORN, OHIO.

In Vol. I, Chap. XIII, of the State Geological Report, the Cincinnati Group or "Blue Limestone" formation is treated of. In that chapter reference is made to the height of the top of said formation above low water mark at Cincinnati at several points, which are indicated on the map opposite page 413 in the volume mentioned. The altitudes of said points are also given in a tabular statement on page 414 of the same volume. Comparing the altitude of said points, the top of said formation is shown to decline four feet to the mile, going north along the arch of the Cincinnati Anticlinal. (See Vol. I, pages 414 and 415). Therein, Station K, one mile above Osborn, is given as having an altitude of 415 feet above low water mark at Cincinnati, whereas, it should have a height of 450 feet to be in harmony with the altitude of the other points given.

Station L, somewhat east but only about four miles south, has an altitude given of 466 feet, and Station F, a very little west and about 13 miles farther south, is given as having a height of 503 feet. Deducting four feet to the mile, in either case, leaves about 450 feet for the height of Station K, instead of 415 feet as given in the survey.

On page 414 of the same Report, Prof. Orton attempts to account for this discrepancy by supposing the whole outlier of the above Clinton Limestone to be there depressed below its normal level; but if any one will take the pains to examine he will undoubtedly come to the conclusion that the Professor has made a mistake in taking the altitude of the top of the Blue Limestone formation at Station K instead of there being a depression of the Clinton Limestone. For, mark, in Volume Second of the Geological Survey Prof. Orton himself, in his article on Greene county, page 662, gives the altitude of Osborn as 410 feet above low water mark at Cincinnati. Now, running up the railroad track one mile to Station K, the elevation of the track at said Station would be about 415 feet above low water mark, and it will be evident to any one on examination that the Cincinnati formation extends to a line of springs 25 to 35 feet higher up the hill. At least the Blue Limestone formation can be seen 15 feet higher than the railroad track, where a wagon road has been cut on the hillside, and the evidences would show its elevation to be about 450 feet instead of 415 feet as given by Prof. Orton. Should this prove to be the case, as it undoubtedly will, the Blue Limestone formation at Station K will be in harmony with the altitude of the same formation at other points given on page 414 of Volume First of the Survey.

REPORT OF THE SURVEY OF ANCIENT EARTHWORKS NEAR OSBORN, O.

BY PROF. J. E. WERREN, URBANA UNIVERSITY.

The hill upon which the earthworks are situated, lies about one mile east of Osborn, southeast of the two R. R. lines (the C. S. & C., and the A. & G. W.); it is of an irregular oval shape, with the greatest axis nearly N. and S., and about one mile in length. It rises from the ancient bed of the Great Miami river, above whose waters it probably projected as an island, 50 feet high (its present elevation from the surrounding valley). The insular character is well attested by the absence of sand and gravel (as well as drift) on the top, and by the sedimentary formation in the outliers south and east, where clay, gravel, and boulders are of common occurrence.

The surface-layer of rocks belongs to the Clinton Limestone, whose base, "or rather the summit of the underlying Cincinnati Group, is a notable water bearer, or is shown by the fine line of springs that issue from this horizon,"* noticeable especially when ascending from the railroad. Other evidences of the Clinton Limestone, especially this, that at many points the beds are extremely friable, are exhibited on this hill in the vicinity of the Springfield and Dayton pike, where one quite remarkable "sinkhole" may be observed; it is without water, and to its lowest bottom studded with lofty trees.†

In the absence of a name given to the hill collectively, the custom is here followed of designating the localities on it, where aboriginal remains are found, by the names of the respective owners, according to which usage the northern extremity (marked A on the accompanying plan) is known as Haddix Hill; south eastern extremity (marked B) is owned by Mr. Mitman, and this portion is known as Mitman's Hill. (Haddix Hill lies nearly entirely in Clarke county; the remainder of the hill, together with Mitman's Hill, is in Greene Co).

*For further information see Part II, p. 666 Ohio Geology.
†"Report of the General Survey of Ohio Geology."

The chief earthworks are found on Haddix Hill, and may be classed in two groups, viz : *Mounds* and *Lines* of *Earthworks*. Of the

MOUNDS

there are three (3) constructed of earth and one stone mound, all on Haddix Hill, and are shown on the accompanying plan marked a, b, c, and d, respectively.

Mound (c), situated in cultivated land, is by far the most extensive, and bears no traces of ever having been disturbed by explorers. Like all mounds of exposed situation it has greatly changed its form, having extended its circumference at the expense of its height; it is now about ten (10) feet high and about sixty (60) feet in diameter, with a nearly circular base.

Mound (a) is the next in extent, about ten (10) feet high and fifty (50) feet in diameter, also circular at the base. It has been opened and explored to some degree; but the investigation, having been carried neither with the necessary amount of carefulness, nor to a sufficient extent, has revealed little beyond the proof that it was a burial mound. The adits still visible warrant the supposition that a future investigation might still be of profit.

Mound (b) is in the line of earthworks "f" and the smallest of the earth mounds, about one-half the size of mound (a). The "Morse Natural History Society of Urbana University" have explored this mound and proved it to be a sepulchral mound. Parts of two skeletons were found in rudely made stone receptacles, overlain by heavy flat stones, serving, evidently, as a covering.

Mound (d) consists of an irregular heap of stones out of whose midst a vigorous tree has grown which is standing now (1878). Were it not for relics, and human remains found there, one would hardly suspect in it the work of a prehistoric race, so much does it resemble an ordinary heap of refuse rock. It is perhaps two (2) feet high and twelve (12) to fifteen (15) feet across.* During the summer of 1877 there were found in this mound a stone pipe, a *fac simile* of which is found on Plate I, (Fig. 2), and a needle. The pipe is quite a curiosity as it has been broken in two, and shows how the owner endeavored to mend the break by cutting a groove in the upper and lower broad surfaces, evidently for the purpose of receiving a wooden or metal crosspiece to hold the fragments together by means

---

*For further notice, see Appendix, "Interview with Esq. Haddix."

of a string, tightly fastened round, which latter the notches on one side prevented from slipping. The pipe is not ornamented. The needle is formed from the tooth of some rodent; the root is perforated to serve as the eye, and the crown brought to a rough point. The original curvature of the tooth seems to be nearly unaltered.

THE LINES OF EARTHWORKS

exist in two groups: farthest to the north runs almost across the entire hill, (east to west) an irregular angular *ridge*, measuring now about ten (10) feet across, two and one-half (2½) feet in height and 1,150 feet in length; it is provided with three sharp angular projections, by the inhabitants of Osborn not unreasonably termed "bastions." They are situated one each near the extremities, and the third about in the middle.* The top of the hill having been used for fifty years as a pasture, the ridge has in many places been trodden down and somewhat lost its distinctness, especially the extremities and the sharp angles of the "bastions."

Immediately behind the ridge (to the south of it) and closely following its course, is a *ditch*, averaging about one and a half feet in depth, and width nearly the same as that of the ridge. Whether this ditch was merely the result of the excavation, whence the material for the ridge was obtained, or in itself intended as a part of the earthwork (fortification) is rather conjectural, yet its regular occurrence along the *same* side of the ridge, and its entire absence from the second line of earthworks, seem to argue rather in favor of the latter theory.

The eastern part of this line of works (e) is lost in the slope of the hill.†

About five hundred feet to the south of earthwork (e), just described, is found another more *regular line of ridges*, (f) showing no distinct ditches accompanying it, and consists of two rectangular enclosures. The larger of them is a field containing a little over 7½ acres. The ridge, whose dimensions are similar to those of the first mentioned, is not everywhere so distinctly traceable as the former, having still more suffered from the destructive causes above mention-

*See table of measurements appended.

†In various places, at irregular distances, the ditch is strewn with stones of various sizes, and as they occur mostly at or near the projections, there are many who believe that these stones had been deposited there by the aborigines for purposes of defense, and to conjecture even that apparatus for throwing them mechanically at the enemy had been in vogue; but as the surface layer of soil is a mere scum it may, with a greater degree of probability, be surmised that these stones were simply, like the earth, thrown up for a like purpose as itself.

ed. The opposite sides are nearly parallel, and the angles very near right angles. The bearings of the longest lines is about south, 15° east.* The enclosure of this field is complete, excepting the middle of its north wall, which affords an entrance of about 18 feet width.

The second rectangular enclosure lies north, projecting from the middle of the former. The walls of this are double, in parallel lines, whose lateral termini are uncertain. The outer wall, or ridge, is considerably lower than the inner one.†

Since the farmer's ax is gradually clearing away the forest trees from the top of the hill, and the cultivated land encroaching more and more upon these grounds of ancient relics, it is fair to presume that the dawn of a new century will hardly recognize the traces of these aboriginal works. These same aggressive causes may have rendered futile the efforts of the party to find some further earthworks mentioned by Esq. Haddix.‡

It is certainly no uninteresting question to inquire here how these aboriginal tribes, with means we generally suppose too crude, were enabled to form with such regularity their angular (chiefly the right angled) figures, and to draw with so much accuracy their parallel lines of considerable length. In connection with this it might be stated that some hold quite confidently the idea, that the direction of principal lines, such as point *nearly* east and west, or north and south, may, at the time of erection, have been *due* east and west, north and south lines, and that by comparison of these with the directions of ancient astronomical structures (or the Egyptian pyramids), whose age is known, the age of this extinct race of builders might be approximately calculated.

On a plat made by a resident of Osborn, Judge Chas. W. Dewey, (and quite carefully executed, and which has been used for reference and comparison, and also depended upon for a general outline of the hill), the walls of this outer earthwork are shown as partially closed; but it was deemed more in accordance with the facts, to leave them undetermined on the plan, even in the face of the ingenious imagination of many which proposes that the outer wall afforded an entrance, at the southeast junction with the larger enclosure, thence

*Notes of the survey at the end of this article.

†For measurements see table of survey at the end of this article.

‡See appendix.

continued uninterruptedly to join it again at the southwest termini, whereas the inner wall was perfectly joined at the southeast junction, but itself afforded a entrance at the southwest junction with the enclosure of the larger field. The two outer ridges would, according to this surmise, have formed a kind of a labyrinthian or masked entrance of this shape:

It has also been deemed advisable, for the present at least, not to indicate a supposed mound in the middle of the outer enclosure where a rise of about a foot is noticed but which seems rather a product of nature.

On a promontory of the southeast extremity "B," upon what is known as "Mitman's Hill," little beyond the Springfield and Dayton turnpike (marked on the plan "p"), there are found *four* nearly *circular depressions*, each occupying a projection of the hill, separated from each other by inlets, which, farther down, are quite steep ravines. Each of these depressions shows a little rise in the center, perhaps produced by the accumulation of the soil, thrown in from the circumference. On the whole these four depressions seem of a more rude nature than the earthworks on Haddix Hill.

In conclusion it might be mentioned, that in and around Osborn a considerable number of mounds and tumuli is found. They are, for the greater part, situated in cultivated fields, and hence through plowing down greatly reduced, altered in shape and less imposing, though some are, notwithstanding these unfavorable circumstances, of still imposing dimensions, as the one on Kauffman's farm (about one mile beyond Mad river), which is still fully fifteen (15) feet high and eighty (80) feet in diameter.

Five miles up Mad river, at Enon, is again a large mound on a prominent hill, and others still farther up along the river bed crown the most important hills, where they may also have served as outlooks and signal stations, for the friendly intercommunication, or warnings and signs of distress, perhaps, among the lost and enigmatic tribes of our now modern country.

The peculiar form of the solitary earthworks described above singularly coincides with the following description of aboriginal remains, mentioned in the "Report of a Geological Reconnoissance, made in

1835 from the Seat of Government, by the Way of Green Bay and the Wisconsin Territory to the Coteau de Prairie, by G. W. Featherstonhaugh, U. S. Geologist." (Doc. 333, printed by order of the Senate) p p 129—132; where the author states :

Having a copy of Carver's Travels with me, an d having always found his descriptions deserving of very great confidence, I had been anxious to discover a remarkable locality he speaks of,* and which, from the doubts expressed by other travellers,† they evidently had never seen. The passage in Carver is so minutely descriptive, and the existence of the remains of a work capacious enough to hold 5,000 men was something so remarkable. that I was solicitous not to miss the place, however troublesome the search, since he does not say on which bank of the river it is, and merely speaks of it as "some miles below Lake Pepin."

On climbing the bank where these evergreen trees were, which is the right bank of the Mississippi, about eight miles S. E. of Roque's‡ trading-house, near the entrance of Lake Pepin, I found myself on an extensive and beautifully smooth prairie. At a distance not exceeding two miles, I saw some unusual elevations to the south ; and, hoping I had had the good fortune to find, at length, the true place, I walked to them, and, on reaching them, was at once persuaded that I had found the locality described by Carver, and which was sufficiently remarkable to justify the description he had given of it. The elevation had the appearance of an ancient military work in ruins ; externally there was the appearance of a ditch, in places filled up with the blowing sand, and having a slope coming down from what might be supposed the walls of the work to the ditch, of about twenty yards. Inside was a great cavity, with irregular

* "One day, having landed on the shore of the Mississippi, some miles below Lake Pepin, whilst my attendants were preparing my dinner, I walked out to take a view of the adjacent country. I had not proceeded far before I came to a fine, level, open plain, on which I perceived, at a little distance, a partial elevation, that had the appearance of an intrenchment. On a nearer inspection I had greater reason to suppose that it had really been intended for this many centuries ago. Notwithstanding it was now covered with grass, I could plainly discern that it had once been a breastwork of about four feet in height, extending the best part of a mile, and sufficiently capacious to cover five thousand men. Its form was somewhat circular, and its flanks reached to the river. Though much defaced by time, every angle was distinguishable, and appeared as regular, and fashioned with as much military skill, as if planned by Vauban himself. The ditch was not visible, but I thought, on examining more curiously, that I could perceive there certainly had been one. From its situation, also, I am convinced that it must have been designed for this purpose. It fronted the country, and the rear was covered by the river, nor was there any rising ground for a considerable way that commanded ; a few straggling oaks were alone to be seen near it. In many places small tracks were worn across it by the feet of the elks and deer, and from the depth of the bed of earth by which it was covered, I was able to draw certain conclusions of its great antiquity. I examined all the angles and every part with great attention, and have often blamed myself since for not encamping on the spot, and drawing an exact plan of it. To show that this description is not the offspring of a heated imagination, or the chimerical tale of a mistaken traveller, I find on inquiry since my return, that Mons. St. Pierre and several traders have, at different times, taken notice of similar appearances, on which they have formed the same conjectures, but without examining them so minutely as I did. How a work of this kind could exist in a country that has hitherto (according to the generally received opinion) been the seat of war to nutitored Indians alone, whose whole stock of military knowledge has only, till within two centuries, amounted to drawing the bow, and whose only breastwork even at present is the thicket, I know not. I have given as exact an account as possible of this singular appearance, and leave to future explorers of these distant regions to discover whether it is a production of nature or art."—Travels through the interior parts of North America, in the years 1766, 1767, 1768, by J. Carver, Esq. Page 57, 58. London, 1778.

†Keating's Narrative, &c. vol. 1, page 276.

‡A half-breed known in the Indian country by the name of Wahjustahchay or Strawberry.

salient angles; and at three different parts were the more regular remains of something like bastions; the cavity was seventy yards in diameter, N. W. and S. E., including the ruins of several terraces; the circumference of this singular place, including the angles, was four hundred and twenty-four yards. Seven hundred yards S. S. E. of this was another, resembling it in form and size; and at an equal distance, E. S. E. from the last, was a larger one, eleven hundred yards round, with similar remains of bastions; this cavity would easily contain one thousand people; its walls, if the word may be applied to them, are lofty, and there is a deep ditch on the south side. In the area to the south I counted six more of these elevations, each having a rude resemblance to the other, with what also appeared to be a line of defence, connecting these works with each other. At the northern end of this singular assemblage of elevations, everything bears the appearance of rude artificial construction; at the southern end, however, and not far from the river, the works pass gradually into an irregular surface, a confused intermixing of cavities and knolls, that might be satisfactorily attributed to the blowing of sand.* There is a growth of oak timber, as Carver observes, upon all this part of the elevations. All the angles and bastions are very much rounded by the weather, and some of the slopes outside consist of sand brought there by the wind. It is undoubtedly true that all the appearances I have described may have been produced by the action of the wind; but those who think so, after personal inspection, are bound to account to themselves why other parts of this prairie, and of other prairies similarly situated, are not blown up, and why the ground covered by these elevations is blown up in such a manner as to resemble artificial works so closely. If, when this curious place becomes more known and investigated, Indian antiquities should be discovered commensurate with the extent of the work, such as the stone instruments and weapons of offence usually found about Indian encampments, it would decide with me the question. If anything of that kind is there, it is probably buried beneath the sands too deep for passing travellers to find. I brought nothing away with me but a plan of the general appearance of the locality, and one or two of the principal elevations.†

*It is sand prairie, covered with a foot or two of vegetable matter.

†We are through the courtesy of the War Department informed that the plans and elevations, mentioned at the close of the above extract, are not on the files of the office of the Chief of Engineers, and regret not to be able to reproduce them here for the benefit of students of American archæology.

EXPLANATION OF THE PLAN AND FIELD NOTES OF THE SURVEY.
A. HADDIX HILL.

a, b, c, Earth Mounds; d, Stone Mound; e, Earthworks, consisting of
I. Ridge, 2½ feet high.

STATION.	COURSE.		CHAINS.	
1 . . .	S. 76° 30′ E.		1.00	
2 . . .	N. 25°	E.	.75	
3 . . .	S. 25°	E.	.50	
4 . . .	East		2.50	
5 . . .	S. 80°	E.	3.34	
6 . . .	N. 70°	E.	.53	
7 . . .	N. 10°	E.	.57	
8 . . .	S. 52°	E.	.52	
9 . . .	N. 62°	E.	.66	
10 . .	N. 35°	E.	1.35	
11 . .	N. 15°	E.	.64	
12 . . .	N. 65°	E.	1.60	
13	N. 55°	E.	.32	
13 to Apex of Bastion			.90	
Apex of Bastion to Station 14			1.00	
14	N. 20°	E.	1.00	
Apex of Bastion Sta. 13 and 14	N. 32°	W.	5.50	To Mound a.

II. Ditch along the Ridge, average 1 to 1½ feet deep.

STATION.	COURSE.		CHAINS.	
(E) 13 .	S. 8°	W.	5.20	To angle (2) in the earthworks to the south.

f. RECTANGULAR ENCLOSURES.

STATION.	COURSE.		CHAINS.	
1	S. 15°	E.	1.05	To terminus.
1	N. 75°	E.	1.75	
2	S. 15°	E.	1.05	
3	S. 15°	E.	.84	
4	S. 75°	W.	.88	To opening.
5 . . .	S. 75°	W.	1.50	
6	S. 75°	W.	2.24	
7	S. 15°	E.	9.50	
8	N. 75°	E.	7.75	
9	N. 15°	W.	9.36	
10	S. 75°	W.	3.00	To Station (4).

B. MITMAN'S HILL.

4 circular depressions, measuring 70, 80, 65 and 80 feet in diameter, respectively.

NOTE.—As the survey of the hill was not the object of the work, its outline on the plan is to be regarded as representing the appearance of the hill in a general manner only.

Plate 9.

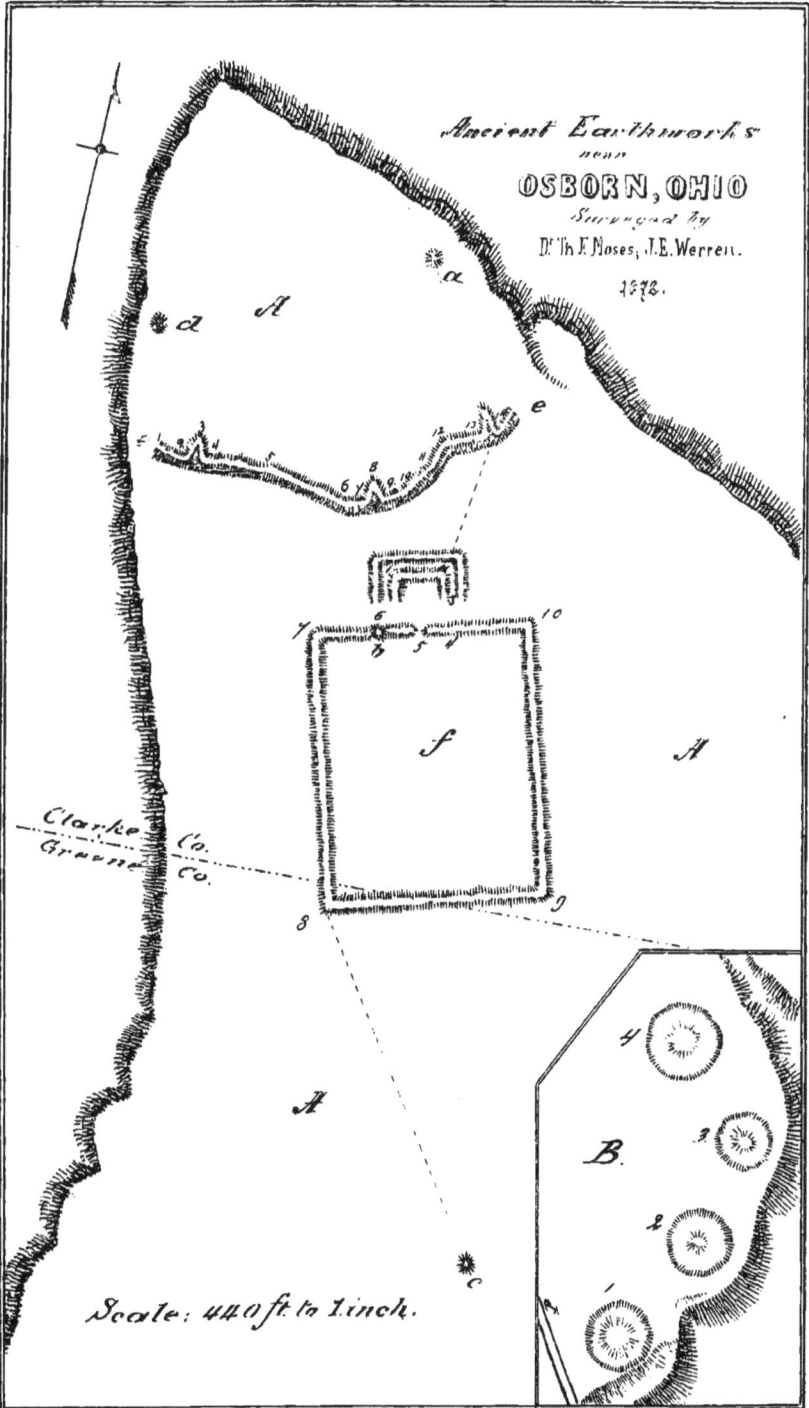

Ancient Earthworks
near
OSBORN, OHIO
Surveyed by
Dr. Th. F. Moses, J. E. Werren.
1878.

Clarke Co.
Greene Co.

Scale: 440 ft. to 1 inch.

J. E. Werren del.

THE HELIOTYPE PRINTING CO 220 DEVONSHIRE ST. BOSTON

APPENDIX.

The gentleman was found at his fireside; he is 87 years old, and his faculties remarkable well preserved. At the age of 11 years he moved with his father to Osborn, where he has been a constant resident for 76 years. The "hill" was his "play house," and upon it Mr. H. discovered the embankment during the first year of his residence at Osborn. He states that numerous wandering Indians of the Shawnee tribe were entirely ignorant of its existence.

The *square mound* (no longer to be found) at the N. W. brow of the hill was described as looking 70 years ago like the foundation of a house, made of earth, raised, and flat.

Mound (a), according to the gentleman's account, was once opened as early as 50 years ago; but nothing extracted beyond some charred bones.

About 40 years ago Mr. H. discovered the stone mound (d) under the following circumstances: he had a wagon made, and in pay for it was digging a well, intending to avail himself of the stones ready in a pile for the construction of the wall. But almost at the very top he struck upon a skeleton: he covered it up again and picked his stones from another place on the hill.

The stone pipe, and needle made of a tooth were found in this same stone mound during the summer of 1877, and generously donated by this gentleman to the Museum of the Central Ohio Scientific Association.

Mr. H. also mentioned the existence of lines of earthworks (no longer traceable) south of mound (c), extending clear across the hill and dividing it into two distinct parts, each division being provided with an excellent descent to the water (which Mr. H. supposed surrounding the hill then), or to springs, and bearing evidence, he thought, that two "camps" inhabited the hill.

Mr. H. tells us that wrestling was quite a regular sport among the young Shawnees, and practiced in a particular manner. He engaged himself in the exercise and practically demonstrated how it was done. The "hold" was in this wise. The right hand passes under the opponents left and firmly grasps the belt (or top of clout) behind. The opponent does likewise. Then the left hand joins the opponent's left in front of the chest: The object then is to raise the opponent with hand or hip, or both, from the ground and to throw him. This was the extent of Squire H.'s information.

Report of a Finding of Pottery on Fox River, near Green Bay, Wis.

BY REV. GEORGE GIBSON, PRESIDENT OF THE ARCHÆOLOGICAL
SOCIETY OF NORTHERN WISCONSIN.

About the middle of June, 1877, during a walk along the track of the Wisconsin Central R. R., and within a mile of the city of Green Bay, I was accosted by the section boss having charge of that portion of the line, who informed me that he had just struck some curious pieces of pottery. Having often ridden with him in my expeditions and shown him fragments of different kinds, he was, in a measure, prepared to recognize anything of the kind when it came under his notice. He was engaged at the time with a gang of men ballasting up the track, and only a few moments before had rescued the remains I found there, which he had deposited on the bank above the pit in which the men were digging; and although pretty well broken up I saw that he had rescued portions of what had belonged to two well preserved vases. The workmen from whose shovels these pieces had been saved, had evidently struck the vessels about midships, and carted away the lower half, which they had rammed into smaller fragments under the ties before my coadjutor himself came up to the spot. Had either of us been there half an hour sooner we might have secured two fine specimens of burial urns. I secured about three-quarters of the rim of one of the vases, averaging four inches in width, together with fragments of the body enough to compose $\frac{1}{2}$ to $\frac{1}{3}$ of the jar. One-half the rim of the second was found with an average depth of two inches. Few or no fragments could be discovered beside the above. These portions of rim I have restored. (See Plate 10).

There were two graves to which the vases belonged, lying east and west. A number of human bones, together with a quantity of fish and bird bones, were found. The human remains were portions of the shoulder, arm, and leg bones. These have been submitted to Dr.

C. R. Brett, of Green Bay, a member of the Archæological Society of Northern Wisconsin. The vases are dissimilar as to construction. No. 1 is globular in form and of about two gallons capacity. Externally the rim is a ¾ in. bevel, ornamented with a series of slanting lines, ⅛ in. in width and cross-barred, resembling the section of a honeycomb cut vertically; or much as though made by the little cross-barred rollers with which our mothers used to ornament the upper crust of the apple pies of old. The inner edge of the rim is tipped off in the same manner. Under the bevel, or flange, is another but shorter series of indentations slightly inclined and cross-barred after the same fashion. The body of the vase is covered with an irregular sort of ropy tracery, running from top to bottom, and looks as though the plastic surface of the vessel, when green had been mopped over with a swab of loose cordage. No. 2 (Plate 10, Fig. 2) has an hexagonal top with globular base. The rim is an inch flange, with a number of creases parallel to the opening running around. The creasing is rather irregularly done, and is crossed at short intervals by slanting lines. The inner edge of the rim, and also the space under the flange, are indented similarly to No. 1, but not cross-barred. Its capacity is about 1 gal. The material of the vases is an equal mixture of red clay and hornblende. Their thickness varies from ¼ to ½ of an inch.

Both vases appear to have been in a state of good preservation, up to the moment they were so unreservedly demolished by the diggers' pick. The fresh surface where broken looks as bright and clean as the fracture in a brand new jug, and the reticulations or cross-bars referred to about the rim and flange are sharp and well defined as when they came well baked out of the maker's oven.

Sta. 2, Fig. 3, shows the place where the vases were discovered. The upper line is intended to represent the ridge running parallel to Fox river. This formation rises in height from 50 to 100 ft. above the Fox. It is a huge mass of clay interspersed here and there with beds of gravel. The latter material seems to have been distributed in the most eccentric way in the mass of hard clay and then covered again by the same material. This ridge is a tongue of land running between two considerable streams, the Fox and East rivers, and is such a point as the Indian is noted for selecting for burial places. Many relics of a people passed away have been discovered at various times, but nothing has been preserved.

In August last I secured at Station 1, ¼ mile north of Station 2, two very perfect copper spear heads and an awl, the latter weighing 7¾ ounces. Here are the remains of a people who certainly existed long before any European ever visited these American shores to tell the story. Station 3, Plate 10, is one mile south of Station 2, and adjoining an old military post, known as Camp Smith in the early days of Green Bay. Here in a corn field during the summer of '76, I found a piece of ground comprising about 2 acres strewn with fragments of pottery, varying from minute particles to a couple of inches in size. Many pieces were nearly decomposed so that they would crumble into dust at the touch. I gathered some six quarts of these fragments which I preserved. From time to time I assorted them into two classes as to difference in ornamentation and form of rim. Selecting according to ornamentation I found about 40 different vases represented that could be definitely re-produced. Taking 50 other fragments more or less mutilated, but still definite as to their relative place in the vase, that is in the rim, we have no less than ninety or a hundred separate vases here represented. Taking these facts into account, together with the condition of the ground which has been unmistakably deeply colored by the abundance of this material, I conjecture that the spot may have been a place where the pottery was extensively manufactured. Near by, human remains have been and still are found in abundance. Two fragments suggest the use of the lathe; one of them belonged to a vase that had a series of concentric horizontal bands of ⅛ of an inch running around it, and their regularity could scarcely be produced in any other way. The other had a set of lines and bands so perfectly drawn upon a smooth field that it is impossible to conceive they could be produced by hand. A third fragment found at Station 2 is of the same pattern as No. 1, and while it shows that the pattern No. 1 has been imitated in the case of No. 3, it bears unmistakable evidence of the fact of having been done by hand. The fish and bird bones found at Station 2 show at least a certain relation to the customs of the Indian tribes that come within the range of history. We know very well that it was a common practice of the North American Indians to place in the grave food, clothing, trinkets, etc., for the use of the spirit until it arrived at the happy hunting ground. This practice implies a use of the vase for it is rarely anything is found in a condition to positively indicate this

Plate 10.

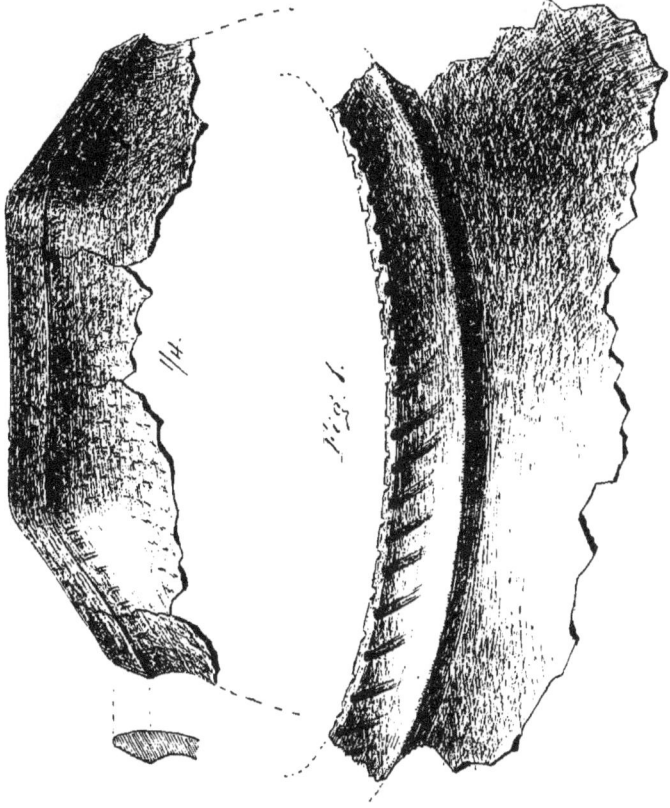

Fig. 2.

Fig. 1.

Fig. 3.

a. *Level of Fox River.*
b. *Wisconsin Central R.R.*
c. *City of Green Bay.*
1. 2. 3. *Stations where relics were found.*

J. E. Warren del.

relation. Obviously the vases would be a receptacle of the food, and not only the food, but the various articles and trinkets that found a place among the donations of friends to the dead.

There would thus be established the fact that the use of vases among the Indians was analogous to that of the burial urn of the Ancients, which not only received the ashes of the departed after the burning, but a variety of articles; rings, precious stones, gold and silver ornaments, utensils for the toilet and other purposes, and in certain cases, as the Hydriataphia informs us, among burial urns dug up in Denmark, instruments of recreation and amusement were discovered.

REPORT OF A SCULPTURED ROCK FROM MARBLEHEAD, O.

BY PROF. J. E. WERREN.

Urbana University has recently come into possession of a queer piece of antiquity consisting of a stone, in size and form like the human body. To the superficial observer the object is hardly more than a stone, but whoever lends the massive stone a few moments' investigation, to him it becomes an object of interest; out of the moss-covered surface first develops a well shaped head of human form, immediately recalling antique sculptures on the rocks in Egypt with their yet unread hieroglyphic mysteries.

The shape of the head is oval, widest across the eyes. The upper part, including the forehead, being not square but rounded, while the lower at the chin is of a sharp oval, adds to the face that stern look which characterizes the appearance of the sphinx of Memphis. The details of the eyes and mouth are suggested by indentations rather than executed in their proper places, while the nose is hardly indicated by a prominence. The ears and arms are mutilated by the marauder's unmerciful strokes, the effects of which are recognized by the rounding moulds left by the hewn out chips. The neck is finely chiseled and in proportion; and the arms, although sadly mutilated, can be traced as crossed over the breast. The body is slightly developed above, but is barely traced out below the clasped hands. The stone out of which this strange image is hewn is a hard limestone of the corniferous group, rich in fossils.

It is perhaps not so much the figure in itself which claims our attention, as the untold history suggested by its strange outlines. What purpose did it serve? When was its form chiseled out of the rock? These are questions which naturally follow the first inquiry, and although the answer can not be gathered from any written or printed chronicle, we will endeavor to present a few remarks on these points as communicated by old settlers living near where the image was found.

Plate II.

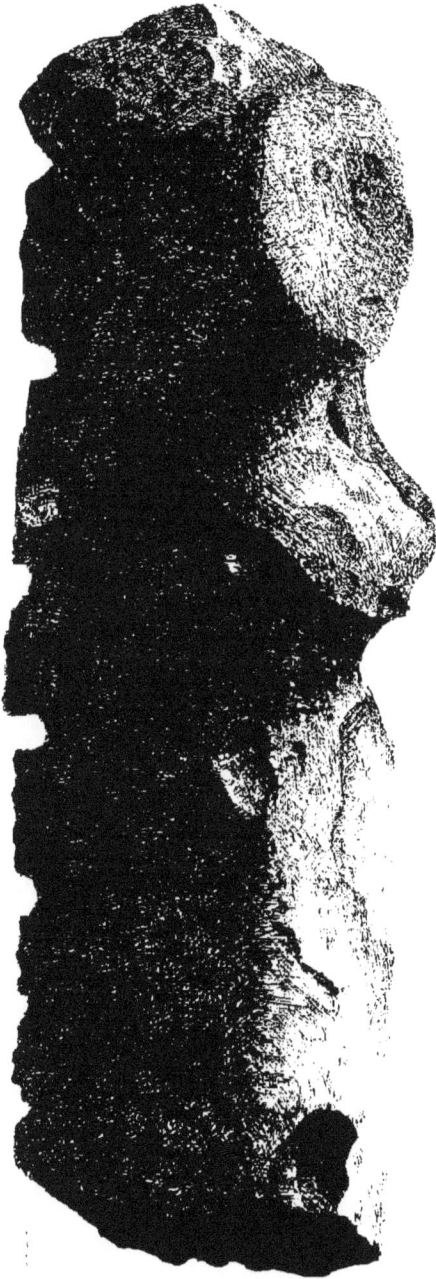

SCULPTURED ROCK FROM MARBLEHEAD OHIO.

J. E. Warren del.

THE HELIOTYPE PRINTING CO. 220 DEVONSHIRE ST BOSTON

The attention of visitors to Ottawa county, O., the greater part of which forms a peninsula in Lake Erie, is soon directed to those little traces whose abundance indicates a territory where once the "wild Indian" roved. These evidences are mainly stone implements, such as arrow heads, spear heads, axes and the like. Even a solitary Indian inscription may yet be found on a neighboring island. The peninsula, although now entirely under culture, offers in its present disguise an aspect that reveals its adaptability to savage life. The northern shore is partly abrupt and rocky, and partly consists of very low land which, on the occurrence of south-western gales, is left entirely dry. Here is at one place a submerged Indian burial ground, known to fishermen only, who occasionally happen there when the water is very low. They also sometimes find there copper implements, although these are rather rare. The eastern shore extends as a sandy ridge a little below the surface of the water almost across to the opposite shore of the lake, and forms the boundary line between the lake proper and what is known as Sandusky Bay. The southern shore sinks gradually into the yellowish green waters of the shallow bay. The extreme slopes of the peninsula were formerly covered with trees to the central ridge, which is a level strip of land. This ridge presents the appearance of *prairies*, and is actually called by that name. These prairies, used only as pastures, are meadows whose rich soil covers with a thin layer the underlying rock of limestone.

The position of a piece of land such as this peninsula offered there an excellent resort for game of all kinds, water fowls together with an abundance of fish, and tradition can not be far wrong in making the spot the often contested ground of many an Indian tribe, the last of which has given Ottawa county its name. Among the many particular places bearing Indian names is especially one on the bay shore which, in its translation, is now known as the "Indian Orchard." The earliest settlers of Ottawa county, who tell of many a treacherous encounter with the red men, as early as the end of the last century commenced the cultivation of isolated spots on the peninsula, but did not as yet dare to build themselves permanent habitations; they had, however, abundant opportunity to observe in peaceful times the customs of the Indians. One of the usages which most engaged the attention of the white man was the strange occurrence of a visit annually celebrated by a strange and unknown tribe, to what was first imagined to be an

Indian grave. The place where the tribe camped and offered unknown petitions was the "Indian Orchard." Under three stout trees (not one of which is now standing) growing close together, was found upon examination a human figure, concealed in the shadow of the overhanging branches, hewn in the rock of the here out-cropping limestone. Further research proved that the place was not a burial ground, there being no indication whatever that such had ever existed there, and the only conclusion was that the figure represented a deity still worshipped there after the tribe had abandoned the peninsula. Many a lover of the remains of Indian art has desired to have this rare relic in a more miniature form to add to his collection, but was forced to content himself with a chip of the stone, which he does not fail to point out to visitors and friends as a "piece of the Indian Idol."

Fantastic stories in connection with this image did not fail to arise, as a matter of course, but their truthfulness could not be verified by the writer. It was, for example, asserted that even while the reservation was wholly in possession of white settlers the Indians kept on visiting the "Orchard," covering the idol with tobacco, pouring whisky over it, and leaving a quantity around it for future supply, which was regularly stolen from the *sanctum* by the white man.

Another story more worthy of note is this, that the image was originally in an upright position, and that carelessness and negligence let it fall to the ground, in which position the civilized population first found it. This theory would naturally suggest a quite recent origin of this Indian monument, and although it might perhaps be unique, it would be of little historical value. The erection of a standing figure requires a high mental culture, and considerably more mechanical skill than the working of a natural rock where the position is given and the setting accomplished with the last strokes of the workman. The writer has therefore taken pains to ascertain the truth of this statement, and by unearthing the idol, where it lay, found that it was actually hewn out of the solid rock, the best proof of which was that it could only be removed by means of iron wedges, a process well known to quarrymen.

The original of this "Idol" is the stone above referred to, now in possession of the Cabinet of the Urbana University. It is all the more worthy of preservation as there are but few such in the country, the Smithsonian Institution at Washington being the only place where the writer is aware of having seen anything approaching it in character. The donor of it is Mr. Mallory, the present owner of the Indian Orchard. The figure is life size and executed in fully one-half relief. With the portion of stone quarried off to give it the necessary strength for bearing transportation its weight is nearly one thousand pounds.

SHELL HEAPS ON THE COAST OF MAINE.

BY PROF. T. F. MOSES.

The shell-mounds described in this paper are located on the Damariscotta river in Maine, some twelve miles from its mouth and just above the "Falls," and the site of the towns of New Castle and Damariscotta, between which the river flows. They lie on both banks of the river and are of great extent. The deposits examined were all on the west bank. Deposits of this character have long been known to the natural historian. In Europe these mounds are found on the Scandinavian peninsula, and in Denmark where they are known by the name of Kjœkkenmœrddings or kitchen refuse-heaps. During the past year similar heaps have been discovered and examined in Japan by Prof. Edw. S. Morse, of the University of Tokio. In our own country they are found in the Mississippi valley along the banks of rivers, and more extensively on the sea coast and the banks of streams contiguous to the coast, from Maine to Florida. Those of the interior are composed of various species of fresh water shells, which those of the sea-board are made up of marine shells belonging to species either now existing or known to have formerly existed in the neighborhood. The exploration of these heaps has been made only within comparatively recent years, and their true character ascertained. These explorations were conducted by Chadbourne and Jackson in Maine; by Prof. Chas. Rau in New Jersey; by the late Prof. Jeffries Wyman, of Cambridge, along the coast of New England and Florida; by Prof. Baird of the Smithsonian Institution, who made the matter a subject of special investigation under the direction of the Secretary; by Mr. F. W. Putnam and his assistants of the Peabody Museum of Archæology, Prof. E. S. Morse, Count Pourtales and others. Besides these explorations many others of a more private character have been carried on.

In the spring of 1859 it was my good fortune to be invited, together with Mr. Jno. M. Brown, of Portland, Maine, to accompany Prof. Chadbourne, of Bowdoin College, (now President of Williams College) on a scientific expedition to the shell heaps at Damariscotta. At this time but little was known of the character of the shell heaps, and they were generally regarded as being of natural origin. Indeed, so eminent an authority as the elder Agassiz had pronounced the shell beds at Damariscotta to be a natural deposit; his decision was based, however, upon specimens sent to him, and would have been quite a different one had he examined the beds in situ. It was with the view of ascertaining the real character of these beds'that Prof. Chadbourne was sent to visit and examine them in the spring of 1859, by the Maine Historical Society. The result of his exploration was to settle for the first time the question as to whether they were the result of geological agencies or the works of men. His report was published in the Transactions of the Maine Historical Society, Vol. 6, and I here insert it in full, as furnished me by the Secretary of the Society, Prof. A. S. Packard, on account of its marking an important event in the history of Archæological investigation :

WILLIAMS COLLEGE, May 18, 1859.

JOHN MCKEEN, ESQ.,

DEAR SIR :—On the twentieth of April I visited the beds of oyster shells at Damariscotta, according to your suggestions. I did not have time to visit all the beds in that region, but I believe I examined those that are considered the most important. I have no doubt that the shells examined by me were deposited by men. This I infer: *First*, from the position of the piles of shells; *Second*, from the deposit beneath them; *Third*, from the arrangement of the shells in piles; *Fourth*, from the frequent occurrence of charcoal mixed with the shells, even to the bottom; *Fifth*, from the fact that fires have evidently been built among them, near the bottom, turning a portion of them to lime, which is mingled with charcoal; *Sixth*, from the mixture of other animal remains, as common clams (*mya arenaria*), thick shelled clams (*venus mercenaria*), fragments of birds' bones, of beavers' bones, with their teeth, and sturgeons' plates.

First, The first thing that strikes the observer is the occurrence of the shells in small piles, ten or fifteen feet in diameter, and apparently two or three feet deep. They seem to rest upon the surface, and to have no soil upon them except that formed by their decomposition and the other substances that would naturally collect from fall of leaves, decay of plants, and movement of dust from year to year. We did not have the time to dig through any of these. I give only the impression that I gained by examining them as they now are, and that is, that they were deposited upon the land in its present position.

Second, Where the river has washed away the bank we have a fine opportunity of examining the deposits beneath the shells, and also their line of juncture with that deposit. We find that deposit made up of sand, gravel, and boulders mingled—a diluvial deposit like all the land in the vicinity beyond the shells; and the line of juncture gives the appearance of shells thrown upon dry ground. There was no appearance of wearing or mingling of the sand with the shells, and in one place, where a boulder was upon the surface of the sand, they seemed to rest against it in a way that precluded, in my mind, the action of the water.

Third, Wherever we found a deep section of shells so lately made that the surface had not decomposed, the open appearance of the shells was marked. They were not mingled with fragments of bone or broken shells or with sand, presenting, in this respect, an entirely different appearance from the great deposit of oyster shells by water at the mouth of the St. Mary's river, Georgia, which I had an opportunity of carefully observing two years ago.

Fourth, In these places, in deep sections, we found fragments of charcoal mingled with the shells under conditions that showed conclusively that it could have been deposited there only as the shells were deposited. The coal left with you was taken out in a deep section very near the bottom. So common did we find the coal that I feel confident it can be found there by any careful observer.

Fifth, In one section a dark line was seen near the bottom of the deposit. Perhaps a foot from the bottom, along that dark line, fragments of charcoal were found, and the shells for a few inches underneath were decomposed, as though they had been acted upon by fire; and in this same place were found most of the fragments of bones left in your possession. I have no doubt a fire was built upon the shells when the bed was about one foot in thickness.

Sixth, The fragments of bones left in your possession are to be submitted to any person desirous of examining them. I consider the jaw and teeth of the rodent animal to be those of a beaver. There is certainly one fragment of a bird bone. And I would call especial attention to the manner in which these bones are broken—as though done with some instrument. I can think of no other means by which they could be broken into such fragments.

The large mass of shells might be used as an argument in favor of deposition by water, but if careful examination proves that they were deposited by men, then the great mass only proves the great number of men or the great length of time during which these shells were accumulating. No man can pronounce an intelligent opinion upon them without an examination. From what I had heard I expected to find that they were deposited by water. There may be beds of shells in that region deposited in this way, but I am fully convinced that those examined by me were deposited by men. I would write more at length but I am very much pressed by my duties. Some future day I should be glad to explore those beds more fully.

Very truly yours,

P. A. CHADBOURNE.

The shell-heaps extend along the banks of the river for a distance of nearly half a mile; one small deposit is seen on a small island near the western bank. They vary greatly in size and form, but are mostly continuous with each other, except where separated by the river. Some of the smaller heaps contain a few bushels only of materials while the largest one examined reaches a height of thirty feet. They rest upon the natural surface of the soil, which is composed of clay and gravel and sandy loam, belonging, geologically, to the drift. Granite ledges rise to the surface in the neighboring hills and numerous bowlders lie scattered about, but none are seen on the surface of the shell-beds themselves. A tolerable heavy growth, chiefly of evergreens, covers the surface of the larger mounds. The soil upon them is quite thin, so that a slight stirring of the surface is sufficient to betray the presence of any underlying shell deposit. The water of the river is salt, and the river is, in fact, like many of the so-called rivers on the coast of Maine, simply an inlet from the sea, like the fjords of Norway. The coast of Maine is notched with these narrow fjords into which the salt water flows, often for many miles into the interior. This peculiar character of the coast line is considered by geologists to be due to the direction of the rocky strata. In New England these jut out irregularly with their ends toward the sea, while along the rest of the coast they lie parallel to it.

The shell-beds first examined by our party were those on the western bank. Through the gradual erosion of the bank by the water the eastern edge of the beds has been worn away, so that a longitudinal section is exhibited along the line of the shore. The shells lie very loosely, are remarkably white and friable, being in a state of partial decomposition and readily falling to pieces when handled. The great majority of them are oyster shells, many of them of great size and of a peculiarly long and narrow shape. One of these shells measured 14 inches in length. Mingled with the oyster shells are found those of the common clam (*mya arenaria*) and the quahog (*venus mercenaria*), and occasionally those of the mussel and pecten or scallop shell. The deposits are entirely free from any admixture of soil or debris of any sort, and one is struck with the appearance which a fresh section presents, the clean white wall of shells looking like a kiln of freshly baked porcelain. Another circumstance that strikes the explorer is the extremely loose condition of the shells, even at the base of a deposit of great depth. The shell may be drawn out with the greatest ease from any portion of the bank, and, with a little caution, in an entire state, although readily crumbling if not handled with great care. From all that I can learn the oyster is no longer found in this vicinity, but tradition has it that it still thrived in the river when the early settlers first came.* In digging down from the surface of one of these heaps at a depth of three or four feet fragments of charcoal were found, and here and there a layer of the same substance. Above and below these layers was some times a conglomerate mass of shells, apparently burned to lime by the action of fire. Among the articles found during the first exploration were some bony scales of the sturgeon, a fragment of the jaw and the incisor of a rodent animal. Near the layers of charcoal fragments of bird bones were also found, apparently the relics of a feast. Two and a half feet below the surface was found a piece of pottery. Similar bits of pottery had been found by Mr. R. K. Sewall of Wiscasset in the neighborhood of a former Indian settlement at Eben-e-cook, in Sagadahoc county. The evidence seemed conclusive that these shell-mounds were not extinct oyster beds left exposed by some former uplift of the Atlantic coast, but the work of aboriginal tribes who repaired to this favored region at certain seasons of the year and celebrated their feasts with the delicious bivalve which must have formerly abounded in these waters. That these feasts were held periodically and perhaps at considerable intervals is shown by the condition of the larger deposits, and especially the large one which slopes to the water's edge on the west bank of the river. Here, at intervals of about a foot, occur lines of vegetable mould marking the limit of successive deposits, so that many years must have passed during their accumulation. The accumulation of such vast deposits and the climatic and other changes which led to the disappearance of the shell-fish once so abundant, will long afford matter for interesting inquiry to the archæologist. Several of the mounds were measured, and some

*Mr. Fuller, Curator of the Portland Society of Natural History, informs me that oysters are yet taken in considerable quantity in the Sheepscott river, which is a stream a few miles west of the Damariscotta.

Plate 12.

Fig. 3.

100 ft.

180 ft.

Fig. 1.

SHELL HEAPS on the COAST of MAINE.

Fig. 2.

Fig. 4.

Fig. 5.

J. E. Warren del.

THE HELIOTYPE PRINTING CO. 220 DEVONSHIRE ST. BOSTON.

sketches and diagrams made. The accompanying representation from a sketch by the writer (Plate 12, Fig. 2) will give a general idea of the larger heaps, though by no means a faithful picture of them. One of the deposits as surveyed by Mr. John M. Brown and myself, has the following dimensions (Fig. 3): Shape oval, length 180 feet, breadth 100 feet, depth 6 feet; height of base above high water mark, 4 feet. The top of the loftiest mound is 31 feet above high water mark. It descends abruptly toward the river, and at its base the action of the water has formed a fine shell beach. The shore line of this mound is about 100 feet in extent, but its extent inland we did not satisfactorily determine, its limits being obscured by a heavy growth of pine, fir and beech, with a dense undergrowth. The specimens collected during this first expedition were deposited by President Chadbourne in the Peabody Museum of Archæology, at Cambridge.

During the present year (July, 1878), in company with Mr. Geo. F. Moses, I made a second visit to the shell-heaps at Damariscotta. Nearly twenty years had elapsed since the time of my first visit, and during that time many visits, including one from Prof. Jeffries Wyman, had been made to these deposits. Near the largest bed a lime kiln had been erected and a large excavation in its summit showed where its operations had been carried on. The shells were burned in the kiln in order to reduce them to a condition that would render them suitable for fertilizing purposes. The marks of numerous explorations were every where visible. During this visit we verified the results of the former expedition with President Chadbourne, finding as before charcoal, animal bones and fragments of pottery. The most interesting find on this occasion was the exposure of a stone wall at the base of a large mound on the bank of the river. The bank had been dug into by other parties and a portion of a wall composed of bowlders was here exposed. Removing a mass of the shells we discovered that the wall was regularly laid up in alternate courses, and as far as we could then ascertain, that the shell deposits had accumulated around and above the wall since it was built. Not having time to continue the exploration, we did not ascertain the extent of the wall. On returning to the village I reported the matter to Dr. R. C. Chapman, a gentleman residing in New Castle who has given much time to intelligent research among the shell-heaps, and have since received from him the following letter :

NEW CASTLE, ME., July 10, 1878.

DR. MOSES:—DEAR SIR:—I have uncovered the entire wall of which you saw only a part. The main wall is about nine feet long and four feet high, terminating in short end walls running out at right angles with the main wall. These end walls are about four feet long and slant off from top to base, the whole wall being on about four feet of shells and in this shape:

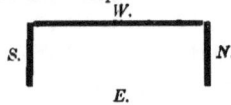

W.

S.   *N.*

E.

Now I think there was a front wall and longer end walls, which have fallen down and rolled into the river or been hauled away. You will see that by supplying what I have supposed belonged to it once that we shall have a very respectable lime kiln, which I have no doubt it was from the fact that I found lime in quite a large quantity; say a bushel or more on what I judge to be the floor of the kiln, it being on a level with the lower stones in the wall. Also a little less than half of a hard burnt brick—like our bricks. No charred wood, bones or implements of any kind were found. It is my opinion that the early settlers of this locality cut a channel into the shell-bank, built this kiln and burned the shells into lime; after they left it the shells fell into it and covered it from view. In process of time the outer wall next the river became exposed by the washing away of the bank and disappeared as I have supposed above. I shall investigate further and will inform you of any new discoveries that I may make.

R. C. CHAPMAN.

Dr. Chapman has gathered an interesting collection of relics from these mounds. Among them are numerous pieces of pottery including quite a large fragment of a gracefully shaped vase which he had restored. A number of stone axes and implements are also in his possession, but these were mostly gathered in the neighborhood of the mounds and not in the mounds themselves. Indeed, there seems to be a remarkable scarcity of such articles in the shell-heaps at Damariscotta. Near the summit of the large mound a human skeleton was exhumed, but this Dr. Chapman regards as an intrusive burial. No other human bones have ever been found in them, the shell mounds here differing in this respect from those on the coast of Florida, where, according to Prof. Wyman, fragments of human bones are often found, evidently the remains of cannibal feasts.

*See article in American Naturalist, Vol. 8, p. 403, entitled Cannibalism.

Incidents Connected with the Early History of Champaign County.

BY JOHN H. YOUNG, ESQ.

The history of the first settlement of Champaign county connects itself very closely with the history of the first settlements of the State.

The first settlements made in what is now the State were made soon after the termination of the Revolutionary war, and were composed largely of Revolutionary soldiers or their families, coming here in the spirit of adventure or driven hither to seek compensation for their services or a home, because of the inability of the General Government to pay them for their services except in lands and land grants.

Indeed there was an express military reservation for the benefit of Revolutionary soldiers which extended into our own county. After the close of the Revolutionary war Virginia ceded to the United States the larger portion of the great domain received by her under charter from King James the First, but in doing so, she reserved all the lands lying between the Little Miami and Scioto rivers, in Ohio, for the purpose of paying the Virginia soldiers who served in the war of the Revolution, and it was distributed amongst the officers and soldiers in quantities proportioned to their several grades in the army. What is known in this county as Ludlow's line was one of the westerly lines of this military reservation.

That grand old soldier of the Revolution, Gen. Putnam, settled at Marietta in the spring of 1788, and this was the first settlement made in the State. It was made at the mouth of the Muskingum river and the settlements continued up the valley of the river. Washington county, of which Marietta became the county seat, was the first county formed in the State.

The next settlements were made near the mouth of the Little Miami river and on the present site of Cincinnati. These settlements were made in November and December of the same year, 1788; and in a very few years thereafter they were extended up the valleys of the

Little and Great Miami rivers to our own beautiful Mad river valley; for history hath it, "that as early as 1795 the region between the Little and Great Miamies, from the Ohio far up toward the source of Mad river, became checkered with farms and abounded in indications of the presence of an active and prosperous population." Hamilton county, of which what is Champaign was part, was the second county established in the State.

How little is known, as it ought to be, of the men who by their heroic struggles with the savage and the wilderness prepared the way for the advent and progress of these settlements—the beginning of a State now so grand in its intelligence, its wealth and its population! Who can properly estimate the value to Kentucky and Ohio of the services of Simon Kenton and Daniel Boone in the conquest of these States from the savage and their allies?

The first settlements seem to have been made upon, and to have followed up, the valleys of the rivers—just as the wild animals and wild Indians had followed and roamed along and dwelt upon them. These favored portions were first sought by the Indian and afterwards by the white man for the same reasons. All up and down the valleys were the great hunting grounds of the Indian.

On the farm immediately west of Urbana, formerly owned by Judge Smith, more recently by his son James C. Smith, have been ploughed up on the fields and found, various Indian implements, such as broad arrow heads, stone pestles, etc. It is said that in the spring of 1795, Tecumseh was established on Deer creek, near the site of Urbana, where he engaged in his favorite amusement of hunting, and remained until the succeeding spring. The Deer creek referred to was probably our town branch, and he was also probably located on the Smith farm through which it ran. There were very fine springs there. On the farm next west of this—the Bryant farm—in a mound on a hill overlooking the Mad river, were found large quantities of human bones. At the mouth of Mac-a-cheek, where it empties into Mad river, are the evidences of an Indian settlement. On the farm now owned by David Miller and Frank McIlvain, in Salem township, was an old Indian corn field. Up the valley of Kings creek, near the town of Mingo, was the village of the Mingo tribe of Indians, to which the great chief Logan belonged—Alfred Johnson now owns the farm embracing that covered by the village, with its great spring—and a beautiful location it is for either Indian or white man. On one of the eminences overlooking Buck creek and the valley, on the farm of John W. Baldwin, have been discovered human bones buried at considerable depth, which from the method with which they were buried and the appearance of the face bones, indicate that they may have belonged to, and probably did belong to, a race of people long anterior to the Indian. Excavations made at considerable depth on the farm of Joseph Townsend, along Kings creek, have revealed human bones that must have belonged to the Indian or an earlier race of people. Like discoveries have also been made on the farm known as the Judge Dallas farm, south of Urbana 3 or 4 miles. Kings creek took its name from the tragic death of an Indian chief upon its bank. He was shot one-fourth of a mile below where Kingston mills now are. It occurred in 1786, during the march of Boone and Kenton with Gen. Logan against the Mad river towns. A portion of the army on horseback, marching up the valley and along near where John Eicholtz now lives, encountered a few Indians. The Indians being headed off from the hills and woods on the east were pursued to the high grass on the bank of the creek, where one of them jumped to his feet, aimed at one of his Kentucky pursuers, but his gun missed fire and the Kentuckian shot and killed him. From his dress and appearance he was supposed to be a King, hence and from that Kings creek took its name.

The Mac-a-cheek towns, ten or twelve miles north of Urbana, were the headquarters of various tribes. It was to these towns that Col. Crawford, the friend and companion of Washington in earlier days, who commanded the unfortunate expedition against the Sandusky towns in 1782, was brought a prisoner, and from which he was afterwards returned to the Sandusky towns and tortured to death with a cruelty so atrocious and fiendish as to excite Washington to tears and stir the hot blood of men everywhere to a desire for revenge.

The State of Ohio was admitted into the Union in 1802. Champaign county was established as a county in the year 1805; and the town of Urbana was laid in the same year, 1805. Montgomery and Greene counties were established in 1803 and each taken from Hamilton and Ross counties. By the Act of the Legislature passed Feb. 20, 1805, the boundaries of Champaign county were fixed as follows: Beginning where the range line between the 8th and 9th ranges between the Great and Little Miami rivers, intersects the eastern boundary of the county of Montgomery, thence east to the eastern boundary of the county of Greene and to continue six miles into the county of Franklin, thence north to the State line, thence west with said line until it intersects the said eastern boundary of the county of Montgomery, thence to the place of beginning. The third section of this Act fixed the temporary seat of Justice at the house of George Fithian in Springfield. The first Court met at the house of George Fithian in Springfield, and its officers were as follows: Francis Dunlevy, President Judge; John Reynolds, Samuel McCollough and John Runyon, Associate Judges; Arthur St. Clair (who was a son of Gov. St. Clair), Prosecuting Attorney; John Dougherty, Sheriff; Joseph C. Vance, Clerk. The first Grand Jury was composed of Joseph Layton, Adam McPherson, Jonathan Daniels, John Humphries, John Reed, Daniel McKinnon, Thomas Davis, William Powell, Justis Jones, Christopher Wood, Caleb Carter, William Chapman, John Clark, John Lafferty, Robert Rennick.

Amongst the first Petit Jury were Paul Huston, Charles Rector, Jacob Minturn, James Reed, James Bishop and Abel Crawford.

Amongst the earliest Attorneys was Thomas Morris, who many years after became United States Senator from Ohio. Ed. W. Pearce was a resident attorney and perhaps the first. Moses B. Corwin, Henry Bacon and James Cooley were amongst the early attorneys. Amongst the Grand Jurors at the May term, 1809, were Frederick Ambrose, Simon Kenton, John Guthridge.

The first trial at the first term of the Court—September, 1805—was the case of the State of Ohio against one Taylor for threatening to burn the barn of Griffith Foos, at Springfield. The form of arraignment was peculiar—the defendant being asked in what manner he would be acquitted, plead not guilty.

At the first session of the Supreme Court, in 1805, the Judges were Samuel Huntingdon, Chief Judge, William Sprigg and Daniel Symmes, Judges. The first case tried was the State against Isaac Brackon, Archibald Dowden and Robert Rennick for assault on an Indian named Kanawa Tuckow. The defendants pleading not guilty and taking issue, for plea, put themselves upon God and their country. They were defended by Joshua Collett, afterwards one of the Supreme Judges, and were acquitted. The jury was composed of William

McDonald, Sampson Talbott, Justis Jones, George Croft and others. A Supreme Court and a jury and the common law! "There were giants in those days."

In the year 1817 Champaign county was shorn of much of her territory, Logan county on the north and Clark county on the south being established in that year. Amongst the first settlers in what is now Champaign county was Gen. Simon Kenton. He was familiar with it and with almost every part of Ohio before he came to settle in it. He first went to Kentucky in the year 1771, when but a boy, and soon became associated with Daniel Boone in his expeditions against and encounters with the Indians in Kentucky and Ohio. In 1778, when on one of his first expeditions through Ohio, he was taken prisoner by the Indians on the north bank of the Ohio river, and lashed on the back of a wild horse, with his hands tied, and a rope around his neck fastened under the horse's tail and around the horse's neck and his feet fastened under the belly, the horse let loose in the woods kicking and plunging through the brush, but escaping from death marvelously. He was then taken to Chillicothe and there compelled to run the gauntlet, and was then brought to the Mac-a-cheek towns and Wapatomica (the latter town was about where Zanesfield, Logan county, now is) and was compelled to run the gauntlet at each place, and was then condemned to a horrible death, but by the generous contrivance of the Mingo Chief, Logan, he was taken on to Detroit and there afterwards escaped and returned to Kentucky. This was in the midst of the Revolutionary war. He was afterwards taken prisoner and again escaped after great sufferings. It is said of him that he was probably in more expeditions against the Indians, encountered greater peril and had more narrow escapes from death than any man of his time. In the year 1786 the Mac-a-cheek towns and other towns at the head of Mad river were destroyed by a body of Kentuckians under General Benjamin Logan. In this attack Col. Daniel Boone and Major (afterwards General) Simon Kenton commanded the advance. He finally settled in Urbana in 1802, and from that time until after the war of 1812, was identified with the interests, the perils and the strifes of the people of this county. He had possessed large quantities of land in Kentucky, but generous and kind hearted as he was brave, he incurred obligations for others which gave him much annoyance. With every opportunity for being rich, the owner of valuable lands in Kentucky and Ohio, he was kept comparatively poor by the assistance rendered to others and the treachery of professed friends.

Judge Burnet, in his notes on the Northwestern Territory, says that he became acquainted with Kenton at Marietta, in the fall of 1796, while attending Court there—Kenton being there as a witness—that he was then possessed of a large estate, and a more generous, kind hearted man did not inhabit the earth. "His residence was in Kentucky, in the vicinity of Washington, where he cultivated a thousand acres of land, equal in fertility to any in the world." Unfortunately he was illiterate and altogether too confiding. He judged others by himself and was not conscious of the imposition to which he was exposed. He believed men were honest, nor did he awake from that delusion till he was defrauded and robbed of his estate.

He had certificates of purchase for 5 tracts of land in Ohio, to-wit: 1,200 or 1,500 acres of land on the Scioto river, also what was known as the Maquechack (Mac-a-cheek) tract, now constituting most of what is the large farm of John Enoch, the tract called Kenton's old place, about half a section, now owned by the heirs of Maj. Wm. Hunt on the road from here to Springfield—he had a cabin on this at one time and lived there; also Kenton's mill tract and a place in possession of one Anderson. The Kenton mill tract is now Lagonda. The tract occupied by Anderson embraced what afterwards became the farms of James Johnson and Orsamus Scott, in Concord Township. Kenton and Col. William Ward owned several tracts together in Ohio, and in the division it was claimed by Kenton as the contract that as he took the Maquechack tract and Ward the tract on which Urbana was located, Ward was to convey him half a section of land adjoining Urbana, which he never did. Col. Ward obtained the patent for Sec. 23 on which Urbana was located, as well as Sec. 22 south of it and Sec. 24 and 25 north of it.

One of the incidents connected with our first Court was the return of the Sheriff on a writ of capias issued against Philip Jarbo and Simon Kenton for the recovery of a debt for which Kenton had become surety. It illustrates the reverence of the Court and officer for the brave Kenton. The return of the Sheriff on the writ was, "found Philip Jarbo and have his body here in Court—found Simon Kenton, but he refuses to be arrested," and he was not arrested. The first jail was on Market street, near where Peter Lawson lives. Kenton was the Jailor about the year 1811, and was at the same time on the jail bounds for a surety debt, and was therefore his own Jailor; but violated neither his duty nor the obligations of his bond.

He took part in the war of 1812. Joined the army of Gen. Harrison and was at the battle of the Moravian towns, where though now an old man, he displayed his usual courage and intrepidity. His skill and courage and knowledge of the Indian character were of great value throughout that contest. He gave his youth, his manhood and his old age to the service of his country. One who knew him and his history well, says "that after he joined the adventurers in Kentucky, about 2 or 3 years before the declaration of American Independence, he was engaged in all the battles and skirmishes between the white inhabitants and the savages." He was also an intrepid leader in most of the expeditions against the Indian towns northwest of the Ohio. Those conflicts continued during the long period of twenty years. He was one of the Judges at the first municipal election held in this town, that first election being in 1816, and he, Anthony Patrick and George Hite, being the Judges. Congress assisted him in his old age by making a generous provision for him. If ever soldier—if ever great and distinguished service to one's country might deserve it, surely Simon Kenton deserved of the people of Ohio some memorial testifying their appreciation of his services and their reverence for his memory, but so far no token or monument marks the place of his burial—he lies

"Unhonored and unsung."

William Owens is supposed to have been the first settler in the county. In the year 1797 he settled on what was afterwards known as Owens creek, about two miles below where Westville now stands. Most of his farm was that on which the late Henry Blose lived. Pierre Dugan, a Canadian Frenchman, some time prior to the year 1800, settled at the head of the prairie east of Urbana, which took its name from him and is known as the Dugan prairie. Some question is raised as to whether he or William Owens was the first settler. His log cabin stood somewhere not far from the homestead of James Long, deceased. Captain Abner Barrett settled on the head waters of Buck creek, about six miles east of Urbana. John Runyon, John Lafferty, Jacob Minturn and Justis Jones settled near and south of where Texas now is, and not long after Henry and Jacob Van Meter.

Nathaniel Cartmill, Benjamin and William Cheney, settled farther down Buck creek valley, near what is now Catawba station, and Parker Sullivan, John Pence, John Taylor, Nathan Fitch, Jacob Pence, Ezekiel Arrowsmith and William Kenton, a brother of Simon, settled along Mad river, west and northwest of Urbana. John Reynolds settled in the western part of what is now Mad river township about the year 1803, and erected the first frame house built in Urbana. This was built on the northeast corner of what is now the Weaver House block. He afterwards built the frame house on the Public Square now occupied by George Collins, and the brick store west of it on the corner of the Square and South Main street, now enlarged and improved.

Arthur Thomas, who was afterwards massacred by the Indians in the war of 1812, settled about five miles north of Urbana. Jacob Johnson and Matthew Stewart settled on Kings creek. John Thomas settled about three miles south of Urbana, about where Mrs. Newell now lives, and had a small distillery up the branch, between where the Newell and Donavan houses now stand. Besides these Felix Rock, John Logan, John Owen, John Dawson and others settled in the county prior to 1805.

The town of Urbana was laid out in 1805 by Col. William Ward, who had entered and owned the soil on which it was laid out. He came from Kentucky, though originally from Greenbrier, Virginia. The first house erected in the town was a log cabin built by Thomas Pearce on market space, immediately north of the present City Hall building and east of South Main street. Christopher McGill, still living in the city, was born in 1802, within the present city limits, in a cabin standing on the east side of the Smith farm. J. H. Patrick, a sturdy pioneer, was born in a log cabin about where Mrs. Keller now lives, in 1811.

Amongst the first settlers in the village were Joseph C. Vance, George Fithian, Samuel McCord, Zeph. Luse, William H. Fyffe, William and John Glenn, Frederick Ambrose, John Reynolds, Simon Kenton and Edward W. Pearce. These first settlers were sterling men and came of good stock. The descendants of most of them still live here. Col. Ward, the proprietor, came from Kentucky, was an intelligent and enterprising man. In Ohio and Kentucky he had some business relations with Kenton, out of which a long continued litigation grew up, involving thousands of dollars in lands and money. He was the grandfather of the distinguished artist, Quincy Ward. He built and lived in a double cabin situated about where Dr. Murdoch now lives. Joseph C. Vance was originally of Virginia stock; served through the Revolutionary war; settled in Washington county, Pennsylvania, where his son, Gov. Joseph Vance, was born; moved to Kentucky, as very many Pennsylvanians and Virginians did, before coming to Ohio; settled in Urbana about or just prior to the year 1805 and became clerk of the first Court and continued clerk until he died, in 1809. His son, Joseph Vance, was elected to the Legislature from Champaign county in 1812. The county then embraced a large territory, the northern boundary being Lake Erie. He was elected to Congress for the first time in 1820 and continued to represent in Congress the district of which Champaign county was part until 1836; was again elected in the fall of 1843 and served for one term. He was elected and served one term as Govenor of the State. I recollect to have noticed him spoken of favorably at one time while he was in Congress, in the old whig times, as candidate for the Vice-Presidency. He was not a learned man, but a man of good strong common sense, had a proper conception of what it was right to do and had the honesty and manliness to try to do it. Samuel McCord became a member of the Ohio Legislature and George Fithian represented the Senatorial District of which Champaign was part. John Reynolds was one of the first Associate Judges; he was a man of great purity of character, of whom it might be justly said, that the town in which he lived and the people by whom he was surrounded, were better and nobler because he lived amongst them. Samuel McCord came from western Pennsylvania, Fithian from Kentucky and Reynolds from Maryland. William H. Fyffe came from Kentucky to Ohio; he filled various county offices acceptibly. Frederick Ambrose became Sheriff and afterwards Treasurer of the county. William and John Glenn were men of great force of character, and if they did not hold offices it was because they did not want them. I think they had great contempt for politics and for politicians. Zepheniah Luse was said to have been acting Commissary during the war. He settled on West Main street about where his son, Col. Luse, now lives. Edward W. Pearce was a reputable lawyer and lived in a house on the lot now owned by Gen. Fyffe's heirs. Of Gen. Simon Kenton I have already spoken.

When Col. Ward laid out the town he dedicated a square in the center of it for public county buildings. A frame house on what is now Court street, standing on the lot next west of Mrs. Keller's was first occupied as a Court House in 1806 and continued to be temporarily occupied until a new Court House was built in the Public Square, in 1814. That new Court House was a spacious brick building, facing to the south. The Court room was on the first floor in the north part of the building; the main entrance to it was from the hall coming in from the south door, on either side of which were the Clerk and Recorder's offices; the other county offices were in the second story. Part of the second story was occupied as a Masonic Lodge, for many of the prominent early settlers were greatly given to Masonry. No fence surrounded the Court House, it was easily accessible; it was the place of all public and political meetings, and for the elections of town and township, and much of the marketing was done on the north side of it; and people from town and county were wont to gather about it and discuss politics and business generally. My first acquaintance with it in a business capacity was as far back as 1832, when as a boy I wrote in the Recorder's office. On the north side of it was a well, from which the water, pure, and bright and cold, was drawn by windlass until in "modern times" that sweet old "moss-covered bucket" was compelled to give way to the unhappy invention of a pump. The old Court House disappeared about 1840 ; the old well remains, however, but was covered from human sight and human taste when the town threw off the rustic but sensible village garb and put on city airs, including oppressive taxes and other evils that style usually begets. My first acquaintance with lawyers practicing in that Court House was in 1835, when I began the study of the law under Israel Hamilton. The lawyers then residing here and practicing were Moses B. Corwin, John H. James, Israel Hamilton, Daniel S. Bell, Samuel V. Baldwin, Richard R. McNemar—Geo. B. Way came soon after. Amongst the distinguished resident lawyers of the past had been James Coolly, who went as charge de affairs to Chili, in 18—, and died there. His widow still lives in this city. He was a man of fine ability. C. P. Holcomb was here for a while and so was Henry Bacon. Hon. Joseph R. Swan was the President Judge, and Ohio has never had a better, a purer, or more learned one.

Gen. Return J. Meigs had his headquarters here in the spring of

1812, preparatory to the arrival of Hull's army to the North, and indeed Urbana became the headquarters for the Northwestern Army. A large portion of the army under Cass, McArthur and Finley, rendezvoused here awaiting the arrival of Col. Miller with the regulars, and occupied the grounds where Thomas Berry and others now reside, until June, when Col. Miller arriving the whole army moved forward for Detroit and Canada, and so favorable were the reports from the army as it reached the neighborhood of the British, and victory seemed so assured for Hull, that the people here raised a tar barrel on a pole in the center of the Public Square and burnt it with many joyful demonstrations. But when the news of Hull's surrender came back over the country none felt the mortification of that defeat more than Urbana.

About the year 1815 a small grist mill was erected where Mr. Fox's woollen factory now stands, but not being very profitable and other mills springing up along Mad river, it was abandoned.

The early settlement of this county, so far as the present boundaries are concerned, was not attended with as many difficulties as the early settlement of many portions of the country. Mad River and its tributaries, King's creek and Buck creek, were rapid streams, and although the country was comparatively level it was easily drained, and so was less affected by miasmatic influences. A large portion of the county, especially Salem township, where were the lands called the barrens and especially the Pretty Prairie, was not heavily timbered; and at a very early period a large part of it was readily opened up as a rich farming country. And whether from prejudice or tradition, or from knowledge derived from those "who have come down to us," I am sure those early settlers were very decent people.

Amongst those "who have come down to us from former generations" and attest of the early history of the town and so link us to the last, are those worthy gentlemen, Judge William Patrick and Col. Douglass Luse.

SIMON KENTON.

A BRIEF DESCRIPTION OF THE MAN AS HE WAS KNOWN IN URBANA.

BY GEO. A. WEAVER, ESQ.

An excellent portrait in existence, a copy of which this Association has, represents Kenton apparently at the age of 70, with a face clean shaven, a kindly expression of eye, a prominent chin, well moulded mouth, long nose, deep overarching eyebrows, and high forehead, somewhat narrow toward the top. The whole face is a striking one, and indicative of a higher type of intelligence than was usual among the frontiersmen even of his time. Unlettered, really unable to read, and quite unused to the ways of civilization, yet he knew the events of his day and the men who took part in the Indian warfare in a way that showed rarer intelligence than is exhibited in most of the registered accounts of those times. Colonel Jno. H. James says that in an interview which he had with him a few years before his death, with the purpose of writing up from his own lips with that peculiar diction of his, his own life, he found him with one of the most remarkable memories he ever met with.

He was a great snuff taker and the habit affected, somewhat, his manner of speech. Not unpleasantly however, for his way of snuffing at the air, with a sidelong hitch of mouth and chin, gave a sort of emphasis or novel energy to his peculiarly abrupt utterances. His manners were usually bland and courteous; and no man, says 'Squire Patrick, was more chaste in habits and conversation. "He resented," says the 'Squire, "the charge that he was a horse thief. 'I never stole horses, except from the hostiles.' He looked upon that as only an act of reprisal. He never utilized the horses but always restored them to the lawful owners or to those who had suffered losses of the kind."

The eccentricity of voice and the archaic style of his language accorded well with the shrewd character of his comments. The very manner of the man would have been worth preserving, and it is most unfortunate that Col. James could not have completed the sketches that he began. Being asked what he thought of General St. Clair, he of fair military fame, but who was singularly unfortunate in his campaign against the Indians in Darke county, he answered, as he twitched his chin and gave one of his habitual snuffs, slowly measuring his words, "Humph! he was a well dis-sip-lined, minister-looking man, but dear me, *he hadn't the brier in his eye.*" As to a writer who wrote more about him than he knew, he said, "That fellow tells a good many lies about me; he wrote about talks I had with Indians, and says I said '*sir*' to 'em," and the old man shook himself and scowled at the reflection. ' "Why, sir, I never '*sirred*' an Indian in my life."

'Squire Patrick relates that he knew him as a quiet, undemonstrative man, whom one might see talking modestly when questioned about himself, walking in the midst of the civilization that grew up about him almost as one who had no share in it. For a year, at one time here, he was kept in prison bounds. That was when the embarrassment of his wealth in Kentucky lands gave him trouble; for some Kentucky creditors had him arrested for debt, under the old barbarous law. The prison bounds in Urbana then extended from Dr. Brown's alley on Scioto street to High street, and from Ward street to Reynolds street. These bounds were afterwards extended to the county limits. He has seen him, he says, walking with his long staff, draw near the same prison bounds, when as if about to pass the line, he would bring up with a sudden halt.

These Kentucky land claims which Kenton has been supposed to have possessed an almost unalienable title to, were partly charged to the account of Col. William Ward, the founder of Urbana, who united with Kenton in early days of Kentucky in their purchase; Ward furnishing some capital and his knowledge of land titles and conveyances. Kenton, to whom was left the business of seeing to payment of taxes, neglected it, involving loss to Ward, who finally closed, by written article, his partnership with Kenton. The consequence was Ward was accused of perfidy and cheating Kenton of his lands; but the fact is Kenton was a shiftless man, and in civilization he did'nt get on well. As to Ward, Kenton scarcely afterward alluded to him, and about these business ventures with him he was always extremely reticent.

In 1811–12 he was jailor here, when the jail, a half log and half frame building, stood on the southeast corner of Locust and Market streets. Jno. McCord, his son-in-law, was with him.

When Hull's army encamped here some of the soldiers seized a friendly Indian whom they falsely accused of killing one of their number. Kenton heard of it, made his way to the scene where the excited company, in spite of all representations and entreaty, were hurrying the unfortunate Indian to the nearest tree. Kenton, roused by honest indignation at the atrocity of their intentions, pushed into their midst. The men fell back before the man who carried in his eye a fierce determination that has often looked terrible to his enemies. He seemed to represent the majesty of the law, who, with his rifle in his hand, placed himself beside the Indian and demanded to know of the mob what they meant by taking an innocent man, and without a word of questioning whether he was innocent or guilty, would put him out of the way? He concluded with the information that the first man who dared to touch a hair of this Indian's head would, the next instant, be a dead man. He then led, without interference, the redskin by the hand outside the town and told him to make tracks.

Results of Meteorological Observations

MADE AT URBANA, OHIO, FOR 25 YEARS; LATITUDE, 40° 6′ NORTH;
LONGITUDE, 83° 43′ WEST; AND 1044 FEET
ABOVE TIDE WATER.

BY MILO G. WILLIAMS.

EXPLANATIONS.

The observations and records were made in accordance with the forms adopted by the Smithsonian Institution; the regular hours of observation being 7 o'clock A. M., 2 P. M., and 9 P. M.

Thermometer.—Besides the regular observations the temperature at sunrise was noticed; this was recorded as the minimum for that day. The annual minimum is the lowest observed point for the year, and the maximum the highest point for that year, without regard to the regular times of observation. The monthly and annual means are obtained from the three regular daily observations.

Barometer.—The height of the mercury is given after the proper reductions are made; and the means are procured from the three regular observations.

Clouds and wind.—The degree of cloudiness is indicated by numbers, the scale being 10 to 0; 10 indicating entire cloudiness, 5 one-half, and 0 entire clearness, &c. The course of the clouds is given to eight points of the compass, and the prevailing course for each day is recorded. The record of the force and course of the wind is entered in the same manner.

TABLE I.—THERMOMETER.

The monthly and annual means; the highest and lowest points each year, and the annual range for 25 years.

	Jan.	Feb.	Mar.	April	May	June	July	Aug.	Sept.	Oct.	Nov.	Dec.	Annual Means.	Lowest Point.	Highest Point.	Annual Range.
1852	19.94	31.38	40.04	48.98	63.22	68.95	74.81	71.65	62.70	58.06	38.70	36.04	51.36	−20	91	111
1853	32.95	32.60	38.66	50.56	60.30	73.30	70.70	71.21	63.98	44.61	45.57	31.40	51.67	−2	92	94
1854	29.46	35.62	43.77	50.96	62.84	70.83	77.51	74.35	76.00	55.27	38.78	30.85	53.36	−4	98	102
1855	25.55	22.53	32.80	53.38	62.63	67.12	73.08	72.50	67.22	49.03	42.05	28.77	50.29	−6	95	101
1856	14.39	19.32	27.31	52.52	58.81	71.73	75.46	66.06	62.10	33.35	42.95	21.93	46.79	−23	97	120
1857	14.37	38.95	34.35	39.56	55.93	67.98	72.84	71.70	65.40	50.00	35.44	35.87	48.53	−19	98	112
1858	36.45	20.70	38.73	41.90	59.00	73.40	73.39	71.83	61.99	55.50	35.36	37.08	51.48	−13	97	110
1859	29.19	32.74	45.60	48.00	66.10	67.93	74.70	71.28	63.16	47.90	33.70	22.00	51.20	−10	93	106
1860	29.11	30.61	42.14	51.62	66.23	69.77	72.73	71.68	60.90	53.30	36.90	26.23	50.95	−11	95	104
1861	27.26	36.71	38.38	49.43	61.67	71.17	74.09	71.34	64.49	50.42	39.88	36.20	51.64	−	93	94
1862	29.03	28.06	37.66	50.93	66.23	66.70	73.35	72.45	64.80	54.21	39.41	35.54	51.12	−1	92	94
1863	32.70	32.95	36.60	50.00	64.57	68.15	74.03	72.61	62.65	47.41	42.67	32.92	51.48	−1	93	92
1864	25.11	30.33	35.73	46.84	62.60	70.23	74.52	72.56	63.12	48.48	41.09	27.71	49.88	−16	95	111
1865	19.96	30.08	30.08	53.07	61.61	74.30	71.81	69.70	72.37	50.00	38.58	31.30	51.37	−15	91	108
1866	25.81	26.27	34.96	55.77	58.73	69.81	75.46	65.44	61.78	53.38	40.86	26.47	49.52	−12	94	109
1867	17.89	31.34	31.68	51.47	55.35	73.25	72.89	73.05	60.46	54.70	45.11	28.57	49.65	−17	91	113
1868	21.75	25.09	42.61	46.66	60.26	69.58	80.48	71.45	60.46	49.39	40.42	25.70	49.49	−	93	92
1869	33.38	32.88	32.00	48.42	59.14	68.32	72.75	73.92	65.20	43.74	34.50	26.93	52.06	1	95	106
1870	29.90	29.07	35.25	53.77	65.56	70.97	76.26	73.13	68.95	54.68	40.38	26.25	52.61	−12	96	113
1871	31.27	33.15	46.85	56.43	63.61	71.41	72.67	74.68	61.54	55.55	37.01	26.25	49.58	−17	94	114
1872	24.13	28.60	31.80	53.74	63.44	71.83	76.81	74.03	66.37	52.18	33.75	20.20	50.05	−18	96	120
1873	22.90	27.25	35.41	49.93	63.40	73.92	78.93	72.95	63.05	48.83	34.12	34.86	52.83	−26	99	110
1874	31.98	32.35	39.38	43.89	65.30	75.20	75.71	71.43	68.94	53.14	39.82	33.65	48.30	−11	94	107
1875	19.08	19.04	35.47	46.77	61.79	68.60	73.37	67.81	61.30	50.01	37.36	38.78	51.16	−14	93	105
1876	36.19	34.77	35.40	50.32	64.23	71.57	75.19	73.13	63.50	49.34	46.53	19.73		−12	93	
Means:	26.51	29.83	37.50	50.14	61.95	70.57	74.47	71.81	60.84	51.52	39.25	29.74	50.86			

TABLE II.—BAROMETER.

The monthly and annual means; the highest and lowest points each year, and the annual range for 25 years.

	Jan.	Feb.	Mar.	April	May	June	July	Aug.	Sept.	Oct.	Nov.	Dec.	Annual Means.	Lowest Point.	Highest Point.	Annual Range.
1852	28.79	28.75	28.75	28.71	28.83	28.89	28.96	28.94	28.95	28.91	28.78	28.77	28.83	27.93	29.47	1.54
1853	28.86	28.76	28.78	28.80	28.86	28.96	28.92	28.91	28.93	28.92	29.00	28.77	28.73	27.94	29.31	1.37
1854	28.84	28.83	28.79	28.84	28.81	28.87	28.99	28.97	28.98	28.93	28.68	28.76	28.86	27.98	29.35	1.36
1855	28.75	28.73	28.73	28.88	28.96	28.78	28.91	28.93	28.94	28.80	28.85	28.78	28.83	27.80	29.43	1.63
1856	28.78	28.65	28.75	28.79	28.77	28.84	28.92	28.83	28.87	28.89	28.81	28.80	28.81	27.97	29.25	1.28
1857	28.87	28.87	28.81	28.70	28.72	28.64	28.89	28.08	29.05	28.95	28.86	28.96	28.86	27.91	29.50	1.69
1858	29.00	28.93	28.96	28.88	28.90	28.96	28.97	29.01	29.00	29.03	29.00	29.00	28.96	28.28	29.49	1.21
1859	29.07	28.92	28.78	28.84	28.79	29.01	29.04	28.57	29.05	29.02	29.92	28.96	28.96	28.16	29.53	1.37
1860	28.93	28.89	28.87	28.81	28.89	29.01	28.84	28.85	28.96	28.89	28.79	28.96	28.87	28.17	29.47	1.30
1861	28.90	28.81	28.82	28.81	28.80	28.81	28.87	28.90	28.96	28.58	28.76	29.01	28.87	28.21	29.43	1.22
1862	28.89	28.85	28.67	28.87	28.80	28.81	28.82	28.90	28.92	28.92	28.89	29.94	28.86	28.12	29.40	1.28
1863	28.85	28.94	28.83	28.82	28.82	28.81	28.85	28.92	28.91	28.91	28.82	28.92	28.87	28.21	29.40	1.29
1864	28.96	28.81	28.72	28.76	28.75	28.90	28.93	28.88	28.85	28.82	28.85	28.80	28.88	28.20	29.44	1.24
1865	28.89	28.91	28.80	28.89	28.79	28.88	28.88	28.98	28.93	28.89	28.90	28.91	28.89	28.13	30.00	1.87
1866	29.01	29.00	28.93	28.83	28.74	28.81	28.89	28.86	28.88	28.89	28.90	28.88	28.86	28.04	29.52	1.48
1867	28.63	28.84	28.87	28.78	28.75	28.83	28.80	28.87	28.95	28.89	28.57	28.86	28.66	28.01	29.49	1.48
1868	28.88	28.80	28.85	28.82	28.71	28.89	28.85	28.90	28.88	28.87	28.01	28.90	28.85	28.10	29.36	1.26
1869	28.84	28.79	28.88	28.79	28.72	28.84	28.85	28.93	28.99	28.90	29.83	28.91	28.83	27.90	29.36	1.46
1870	28.85	28.75	28.78	28.79	28.79	28.80	28.82	28.85	28.93	28.85	28.90	28.87	28.83	28.12	29.34	1.22
1871	28.95	28.82	28.76	28.71	28.64	28.80	28.85	28.83	28.96	28.93	28.88	28.87	28.55	28.17	29.32	1.15
1872	28.88	28.81	28.84	28.84	28.83	28.82	28.86	28.84	28.96	28.92	28.89	28.86	28.84	28.12	29.44	1.30
1873	28.79	28.81	28.83	28.73	28.78	28.82	28.88	28.91	28.90	28.90	28.81	28.93	28.84	28.14	29.44	1.30
1874	28.92	28.90	28.88	28.86	28.78	28.83	28.86	28.94	28.91	28.91	28.95	28.95	28.89	28.11	29.52	1.41
1875	28.99	28.91	28.82	28.81	28.77	27.84	28.88	28.85	28.90	28.85	28.87	28.78	28.86	28.09	29.37	1.28
1876	28.93	28.90	28.80	28.84	23.65	28.77	28.87	28.91	28.83	28.82	28.80	28.87	28.85	28.06	29.44	1.38
Annual Means:	28.89	28.84	28.81	28.80	28.80	28.84	28.89	28.90	28.93	28.90	28.86	28.88	28.86			

TABLE III—WEATHER.

The number of clear, fair and wholly cloudy days; the number of days on which there was rain, snow, or thunder; the quantity of snow and rain in inches; the degree of cloudiness, and the point of the compass from which they came; and the means of each for 25 years.

	Clear.	Fair.	Cloudy.	Rain.	Snow.	Snow Covered Ground.	Quantity of Snow.	Quantity of Rain.	Thunder.	Mean Degree of Cloudiness.	Prevailing Course of Lower Clouds							
											N.	N.E.	E.	S.E.	S.	S.W.	W.	N.W.
1852	32	86	89	123	36	23	33.11	58.84	31	5.65	11	11	13	8	28	47	168	49
1853	30	163	50	91	31	22	28.95	45.20	35	5.32	12	13	9	5	16	60	157	43
1854	50	128	41	110	33	23	27.49	41.35	48	4.85	14	15	9	0	22	68	146	41
1855	35	92	69	129	40	61	45.46	37.47	48	5.72	8	15	4	10	17	72	157	48
1856	38	156	44	86	48	92	39.69	30.87	27	4.77	13	9	10	4	13	60	166	51
1857	14	130	63	115	49	48	32.59	30.77	33	5.68	17	15	4	7	27	67	163	51
1858	21	119	82	120	23	51	31.54	40.99	46	5.64	17	14	9	9	27	66	142	60
1859	26	115	78	122	24	45	31.70	36.51	43	5.61	18	16	7	6	24	67	165	36
1860	22	124	57	123	40	45	27.11	35.72	50	5.71	15	12	13	10	15	52	172	55
1861	22	153	49	115	27	41	17.54	36.35	36	5.27	23	23	17	5	20	63	136	41
1862	37	133	59	97	39	57	41.98	37.79	28	5.35	31	16	18	12	35	63	124	37
1863	37	140	59	92	45	55	47.40	36.56	35	5.64	30	17	11	12	26	52	163	43
1864	29	161	38	108	37	45	35.17	32.19	43	5.22	26	9	11	6	33	76	136	36
1865	32	150	38	111	38	39	20.65	46.04	18	5.54	18	11	12	7	22	70	150	45
1866	33	141	33	119	42	75	22.77	49.62	30	5.57	25	8	14	11	18	69	130	56
1867	34	143	42	98	48	64	59.09	31.86	40	5.36	30	19	13	8	21	70	137	42
1868	29	148	45	104	43	75	38.45	46.31	44	5.48	23	13	14	8	22	87	186	35
1869	27	139	48	102	49	67	55.77	42.71	33	5.72	33	25	14	15	19	56	149	29
1870	35	152	50	84	44	48	43.77	32.30	34	5.25	18	18	7	10	26	64	188	40
1871	44	139	42	99	31	59	22.73	30.64	44	5.20	19	18	9	11	22	74	145	47
1872	21	161	41	84	59	61	45.33	28.53	30	5.31	12	16	5	9	23	87	129	48
1873	42	116	64	97	46	87	37.00	37.17	38	5.36	26	20	12	12	23	67	122	40
1874	43	140	72	75	35	55	26.86	34.03	25	5.13	14	21	8	15	28	69	146	38
1875	32	120	59	109	43	39	30.87	43.16	31	5.61	18	19	12	12	23	76	125	44
1876	32	143	53	99	31		41.53	41.93	45	5.73		15	11	9	28			
Annual Means.	31.4	135.5	54.5	104	39.4	48.5	35.46	39.75	37.5	5.43	19	15	11	9	23	67	146	44

TABLE IV.—RAINFALL AND WIND.

The quantity of rain in inches for each month and year, and the annual means for 25 years; also the mean force of the wind each year, and the number of days the prevailing course of the wind was from eight points of the compass; and the number of days calm.

Year	Jan.	Feb.	Mar.	April	May	June	July	Aug.	Sept.	Oct.	Nov.	Dec.	Whole Quantity	Mean Force	N.	N.E.	E.	S.E.	S.	S.W.	W.	N.W.	Calm
1852	2.74	3.18	4.99	5.69	1.41	4.21	3.68	3.05	6.03	3.59	5.64	11.16	58.84	1.24	14	24	28	34	43	60	99	49	15
1853	1.79	4.01	2.53	4.42	3.06	4.72	4.16	8.44	4.16	2.49	3.67	1.75	45.20	1.63	37	36	16	11	45	73	80	51	16
1854	3.61	3.76	5.41	5.75	6.02	2.80	1.67	1.99	1.97	4.46	2.62	1.29	41.35	1.87	25	26	27	17	44	76	88	45	17
1855	3.97	1.66	3.40	2.56	6.72	10.78	6.17	1.23	8.26	3.18	5.18	8.86	57.47	2.33	11	22	9	41	59	91	78	44	9
1856	1.02	1.90	1.85	1.90	3.84	3.29	3.89	2.37	2.91	2.08	3.62	3.02	30.87	2.20	17	18	18	44	47	84	68	45	15
1857	1.16	3.01	.96	6.41	7.50	3.05	4.23	4.56	1.84	1.75	5.65	2.54	39.77	2.08	11	23	18	46	62	76	73	46	20
1858	2.03	1.48	.96	3.86	7.50	5.26	3.60	4.86	1.97	1.75	3.39	4.90	40.99	1.97	9	23	23	46	59	65	76	47	10
1859	2.30	3.05	4.16	1.25	6.41	3.37	.80	3.35	3.35	1.26	4.69	4.66	36.51	2.24	18	29	13	49	74	69	76	31	6
1860	1.87	1.62	.76	6.30	1.07	3.20	6.21	3.93	2.59	2.00	2.42	3.17	35.72	2.32	18	21	30	40	65	80	77	47	8
1861	1.97	2.95	2.95	3.10	2.53	3.69	4.02	2.29	3.42	2.83	2.83	1.21	36.35	1.89	22	29	19	28	43	67	82	43	11
1862	3.01	2.47	4.35	3.70	3.70	3.20	2.10	2.33	.60	1.13	3.08	4.82	37.79	1.85	32	30	40	35	42	70	72	54	3
1863	6.36	3.18	3.50	1.69	3.54	1.31	.84	5.47	8.13	3.86	3.01	4.31	36.56	1.65	23	19	21	31	65	74	57	42	28
1864	1.89	.55	2.33	2.31	2.21	3.82	4.62	6.66	3.71	1.80	3.53	3.64	32.10	1.99	30	15	25	24	49	86	107	35	14
1865	1.55	1.07	1.88	6.92	4.11	8.82	4.74	2.01	5.32	1.22	.73	3.20	46.04	1.96	18	18	18	21	60	45	86	36	12
1866	3.39	3.25	3.51	1.36	1.59	5.54	4.71	3.57	15.88	1.22	3.27	2.11	49.62	1.72	24	21	20	16	51	82	107	43	23
1867	1.56	3.85	3.08	3.48	2.27	1.08	2.87	2.08	.32	2.01	1.77	4.09	31.86	1.57	34	23	22	26	36	95	89	61	10
1868	2.44	1.08	3.33	3.48	6.19	10.38	1.88	2.63	3.81	1.17	1.77	1.90	46.31	1.81	27	23	17	15	43	77	85	55	12
1869	1.50	3.40	5.73	.43	7.09	2.49	6.52	1.01	3.32	1.89	4.21	3.12	42.71	1.90	30	57	10	26	30	99	107	37	10
1870	6.66	2.42	4.26	1.14	.68	3.07	2.63	2.34	.47	4.00	1.90	2.32	32.30	1.96	21	40	12	21	42	77	112	32	12
1871	1.35	1.85	2.74	2.84	2.00	3.65	2.45	6.48	.25	1.20	3.33	2.30	30.64	2.07	13	25	10	27	82	82	90	56	4
1872	1.21	1.32	1.46	3.28	1.34	1.24	7.83	4.96	.75	1.95	1.12	2.57	28.58	1.81	18	29	18	27	40	124	74	42	5
1873	2.78	1.14	2.42	3.44	3.40	2.83	6.76	1.19	2.49	3.21	1.92	5.59	37.17	1.51	20	32	10	28	119	82	66	38	13
1874	4.68	4.23	2.70	4.11	1.32	2.19	3.24	2.77	1.61	.53	3.19	4.66	34.03	2.08	24	25	10	28	88	119	66	38	10
1875	1.25	2.13	2.61	1.67	3.09	6.69	9.69	3.74	2.54	2.94	3.72	3.09	43.16	2.07	20	32	17	28	42	107	68	34	17
1876	5.62	2.99	4.24	3.04	2.66	2.83	6.71	1.63	3.73	2.51	2.33	1.24	41.93	1.79	24	29	15	27	55	104	69	34	17
Means:	2.71	2.41	3.31	3.48	3.61	4.16	4.18	3.54	3.36	2.34	3.16	3.45	39.75	1.92	21.9	25.1	19.2	28.2	49.7	82.2	82.3	44.0	12.7

TABLE V.

The Return of Migratory Birds, and the Early Blossoming of Trees and Plants, and Mean Times, and the Range of Their Appearance for 20 years.

	Robin.	Blue Bird.	Meadow Lark.	Martin.	Brown Thrush.	Spring Beauty.	Peach.	Cherry.	Pear.	Flowering Almond.	Apple.	Red Bud.	Lilac.	Native Plum.	Native (Crab Apple).
1859	Feb. 25	Feb. 11	Mar. 1			April 1	April 11	April 11	April 19	April 19	April 21	May	May 1	April 30	
1860	Feb. 22	April 26	Mar. 1	April 7		April 9	April 12	April 12	April 18	April 18			May 2	April 22	
1861		Feb. 13	Feb. 27	April 14		April 12	April 22	April 22	May 1	May 1	May 8	May 11	May 11	May 4	May 16
1862	Feb. 27	Mar. 6	Feb. 28	April 13		April 13	April 22	April 23	May 6	May 3	May 6	May 6	May 12	May 10	May 18
1863	Feb. 20	Mar. 29	Feb. 28	April 16		April 24		May 8	May 8	May 7	May 10		May 18	May 10	May 21
1864	Mar. 2	Feb. 24	Feb. 24	April 6		April 7	April 17	April 17	May 7	May 3	April 27	April 24	May 18	May 10	May 21
1865	Mar. 2	Feb. 20	Mar.	April 4	April 11		April 30	May 1	April 24	April 24	April 28	April 27	April 27	April 28	May 15
1866	Feb. 25	Feb. 9	Mar. 9	April 11	April 27	April 17	April 29	April 30	May 6	May 6	May 9	May 9	May 8	April 24	May 12
1867	Feb. 9	Feb. 16	Mar. 12	April 21	April 22	April 22	April 28	April 29	May 4	May 4	May 10	May 7	May 13	April 28	May 28
1868	Feb. 20	Feb. 21	Mar. 9	April 19	April 16	April 22		April 28	May 2	May	May 2	May 9	May 7	May 12	May 20
1869	Mar. 1	Feb. 24	Mar. 12	April 30	April 21		April 9	April 18	May 11	May 11	April 18	May 2	May 5	May 7	May 23
1870	Feb. 25	Feb. 23	Mar. 4	April 11	April 19	April 9	April 28	April 9	April 11	April 11	April 18	April 18	April 22	May 2	May 13
1871	Feb. 24	Feb. 23	Mar. 9	April 13	April 10	April 13	April 28	April 28	April 30	April 30	May 6	May 6	May 6	April 20	May 2
1872	Mar. 2	Feb. 18	Mar. 11	April 9	April 19	May 1		May	April 8	May 10	May 13	May 18	May 18	May 20	May 22
1873	Feb. 20	Feb. 16	Mar. 18	April 30	April 10	April 22	May 7	May 16	May 9	May 10	May 13	May 12	May 12	May 12	
1874	Feb. 24	Feb. 13	Feb. 11	April 13	April 18	April 12	May 16	May	May 9	May 14	May 18	May 19	May 20	May 12	May 14
1875	Feb. 24	Mar. 10	April 8	April 5	April 17	April 18	May 2	April 28	May	May 2	May 8	May 15	May 16	May 9	May 17
1876	Feb. 25		Mar. 6	April 7	April 31	April 2	April 11		May 2	May 2	May 13	May 8	May 16	May 9	May 11
1877	Feb. 24	April	April		April 17								April 24	May 9	May 19
1878	Feb. 19	Mar. 7	Mar. 19		Mar. 31		April 11	April 13	April 14	April 14	April 21	April 21	April 21	April 20	
Means:	Feb. 24	Feb. 23	Mar. 10	April 14	April 17	April 17	April 24	April 27	April 28	May 3	May 5	May 5	May, 8	May, 8	May 17
Range:	21	30	46	26	18	30	38	30	29	27	30	34	28	22	26

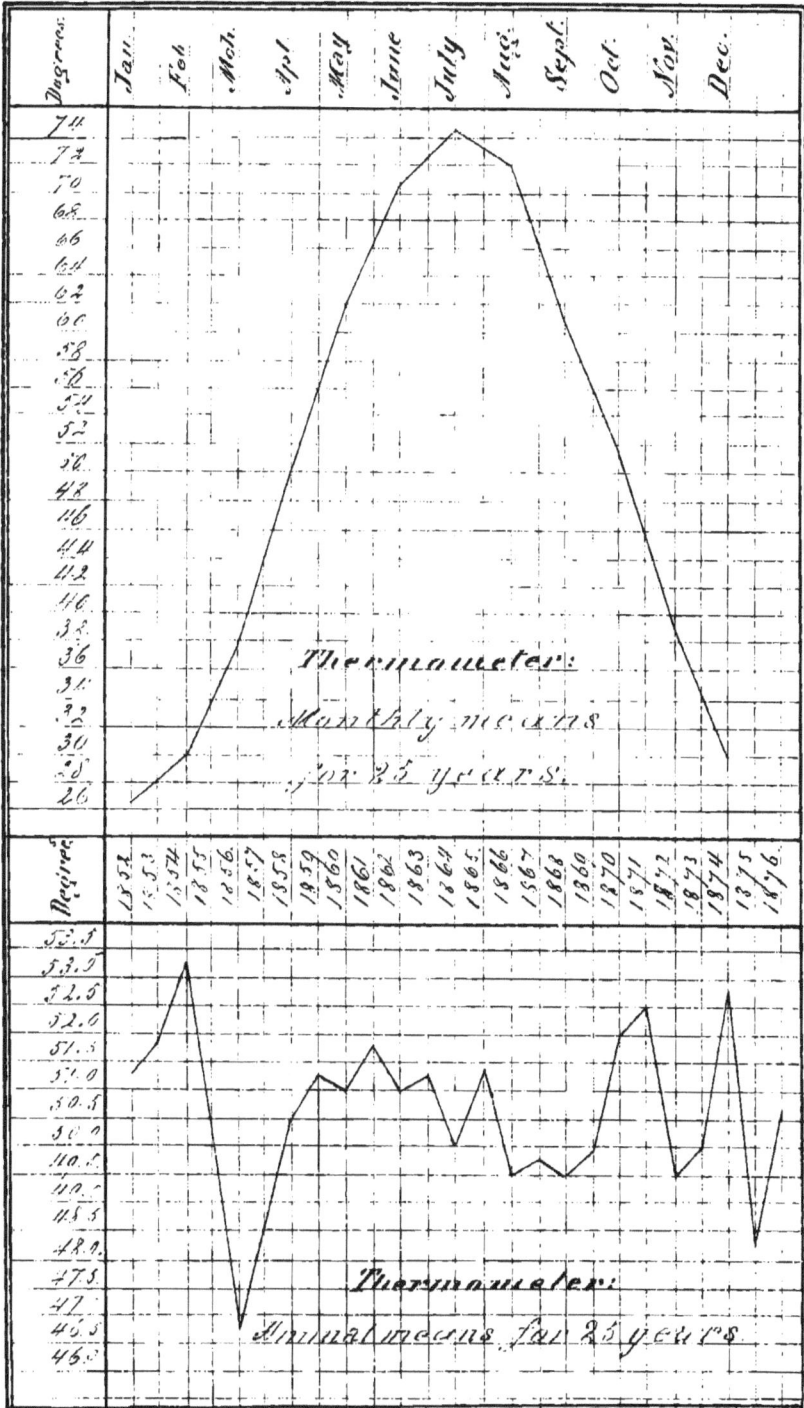

Plate 13.

Degrees.

Jan. | Feb. | Mch. | Apl. | May | June | July | Aug. | Sept. | Oct. | Nov. | Dec.

74
72
70
68
66
64
62
60
58
56
54
52
50
48
46
44
42
40
38
36
34
32
30
28
26

Thermometer:

Monthly means

for 25 years.

Degrees.

1852 | 1853 | 1854 | 1855 | 1856 | 1857 | 1858 | 1859 | 1860 | 1861 | 1862 | 1863 | 1864 | 1865 | 1866 | 1867 | 1868 | 1869 | 1870 | 1871 | 1872 | 1873 | 1874 | 1875 | 1876

54.0
53.5
52.5
52.0
51.5
51.0
50.5
50.0
49.5
49.0
48.5
48.0
47.5
47.0
46.5
46.0

Thermometer:

Annual means for 25 years.

Prepared by, Miss G. Williams.

HELIOTYPE.

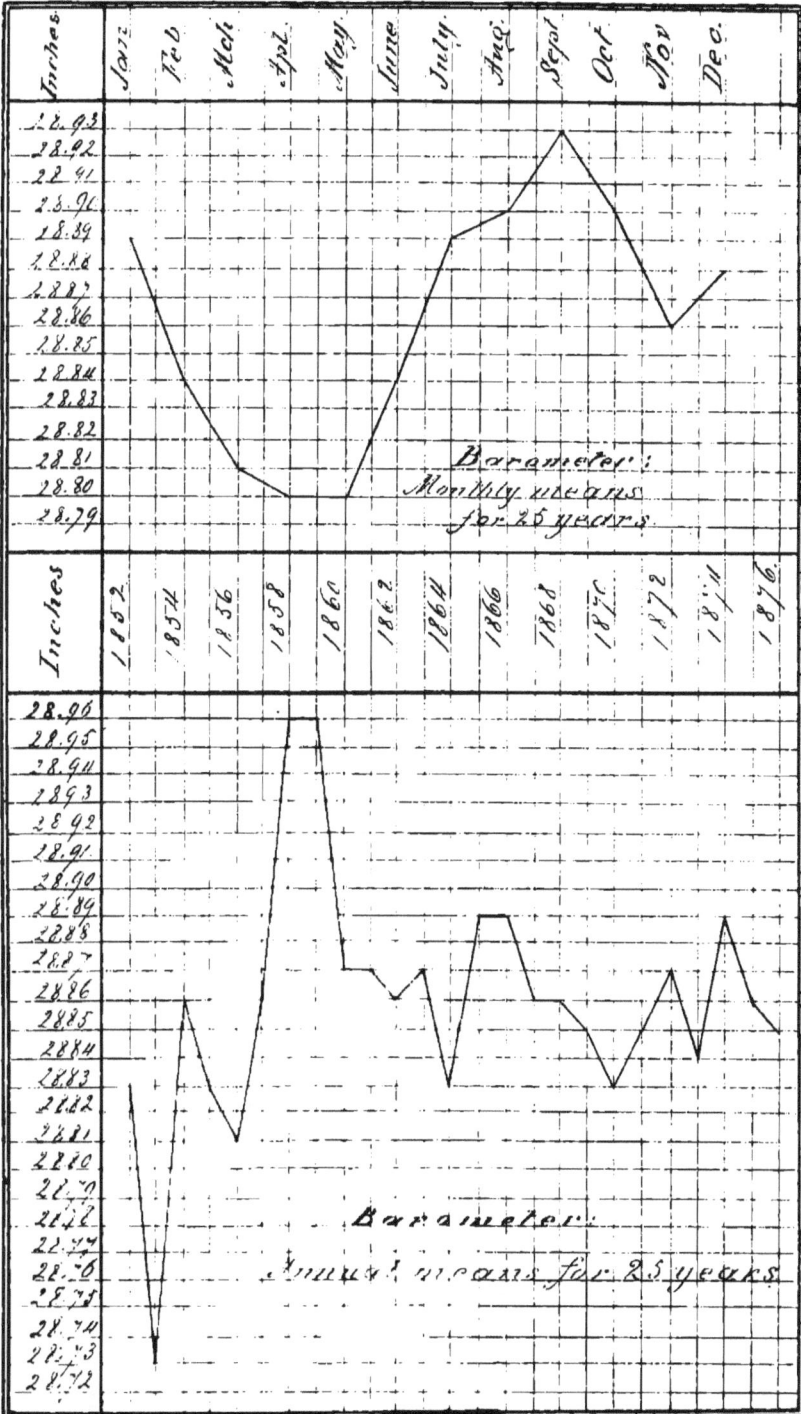

Plate 14.

Inches	Jan.	Feb.	Mch.	Apl.	May.	June	July	Aug.	Sept.	Oct.	Nov.	Dec.

28.93
28.92
28.91
28.90
28.89
28.88
28.87
28.86
28.85
28.84
28.83
28.82
28.81
28.80
28.79

Barometer:
Monthly means
for 25 years

Inches	1852	1854	1856	1858	1860	1862	1864	1866	1868	1870	1872	1874	1876

28.96
28.95
28.94
28.93
28.92
28.91
28.90
28.89
28.88
28.87
28.86
28.85
28.84
28.83
28.82
28.81
28.80
28.79
28.78
28.77
28.76
28.75
28.74
28.73
28.72

Barometer:
Annual means for 25 years

Prepared by H. G. Williams

HELIOTYPE

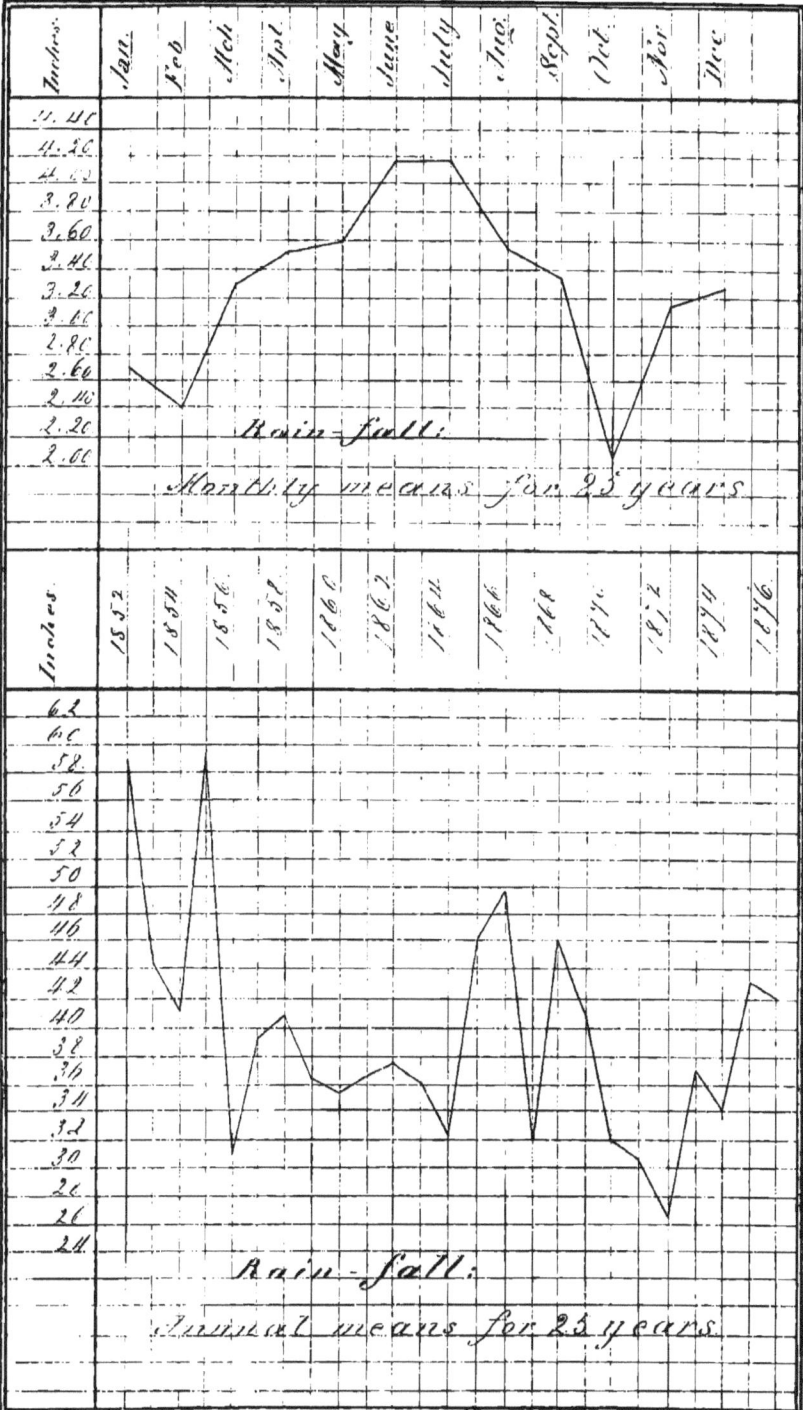

Plate 15.

Inches	Jan.	Feb.	Mch.	Apl.	May.	June	July	Aug.	Sept.	Oct.	Nov.	Dec.	
4.40													
4.20													
4.00													
3.80													
3.60													
3.40													
3.20													
3.00													
2.80													
2.60													
2.40													
2.20													
2.00													

Rain-fall:
Monthly means for 25 years

Inches	1852	1854	1856	1858	1860	1862	1864	1866	1868	1870	1872	1874	1876
62													
60													
58													
56													
54													
52													
50													
48													
46													
44													
42													
40													
38													
36													
34													
32													
30													
28													
26													
24													

Rain-fall:
Annual means for 25 years

Prepared by John G. Williams.   HELIOTYPE.

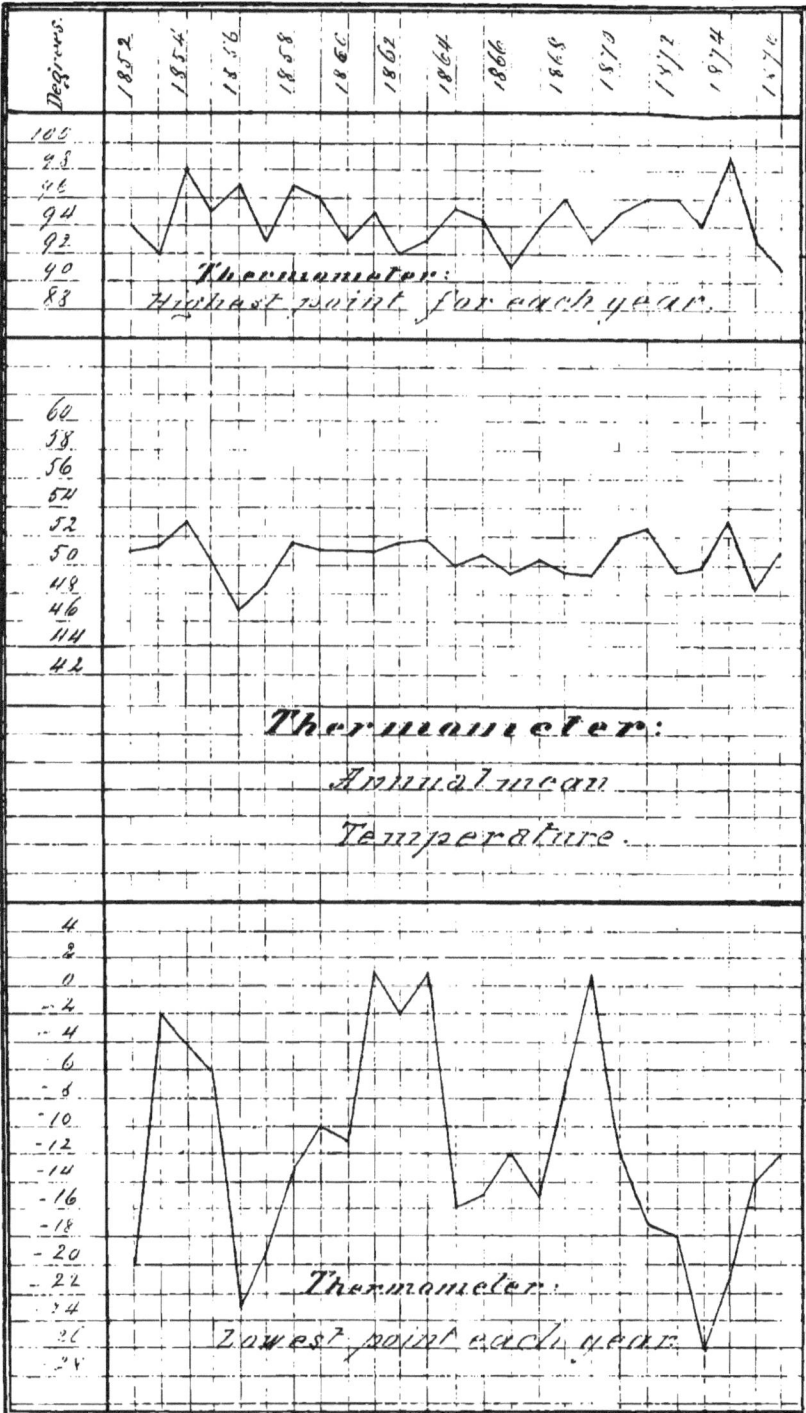

Plate 16.

Degrees.	1852	1854	1856	1858	1860	1862	1864	1866	1868	1870	1872	1874	1876

105
98
96
94
92
90
88

Thermometer:
Highest point for each year.

60
58
56
54
52
50
48
46
44
42

Thermometer:

Annual mean

Temperature.

4
2
0
- 2
- 4
- 6
- 8
- 10
- 12
- 14
- 16
- 18
- 20
- 22
- 24
- 26
- 28

Thermometer:

Lowest point each year.

Prepared by: Milo G. Williams

HELIOTYPE.

APPENDIX.

ADDITIONAL MEMBERS.

ACTIVE MEMBERS.

Wm. A. Brand, Urbana, O.   Jno. B. Niles, Niles, Mich.

HONORARY MEMBERS.

Jno. H. James, Urbana, O.   Lemuel Weaver, Urbana, O.
Wm. Patrick, Urbana, O.   Wm. M Murdoch, Urbana, O.

ADDITIONAL DONATIONS.

EMANUEL STOVER, Urbana—Pestle and Flesher from Madriver township, Champaign county.

HARRY MILLER, Urbana—2 Arrow Heads from Clark county.

THOS. RAWLINGS, Urbana—Stone Pestle.

WM. QUEIN, Urbana—Arrow Head and Borer found in Champaign county.

J. M. POYSELL, Urbana—Stone Ax from Miami county.

R. H. CHEETHAM, Urbana—Water-worn Stone, plowed up on his farm in Champaign county.

DR. L. M. AYERS, Champaign county—Stone Charm, found with a skeleton buried ir sand in a bed of clay, on Treacles Creek; also, 31 Flints and Spear Heads, and 2 Stone Axes from same creek.

PEARL CRAIG, Champaign county—Fossil Coral, 2 specimens.

C. E. COLWELL, Urbana—Snout of Shovel Fish, from Illinois river.

W. A. BRAND, Urbana—Copies of old newspapers, as follows: *Farmer's Watchtower*, Urbana, O., June 9, 1813. The *Ohioan and Mad River Journal*, November 19, 1825, Urbana, O. *Farmer's Friend*, Nov. 8, 1820. Copies of *Mad River Courant*, Urbana, O., Aug. 16, 1828; Oct. 22, 1831; Jan. 7, 1832; Aug. 20, 1831. *Country Collustrator*, Urbana, O., Aug. 18, 1831. *Columbian Herald*, Dec. 20, 1787, published at Charleston, S. C. *Western Statesman*, Columbus, O., Dec. 30, 1826, and April 16, 1827. *Wilson's Knoxville Gazette*, Jan. 6, 1812. The *Log Cabin*, New York and Albany, Aug. 29, 1840. *Western Reserve Chronicle*, Warren, Trumbull county, O., Dec. 28, 1820. *Anti-Slavery Bugle*, Salem, O., Dec. 8, 1848. *Christian Banner*, Fredericksburg, Va., June 11, 1862. Address to Jackson Electors of the 5th Congressional District by Wm. Russell, West Union, Sept. 1832. *Fac Simile* of the last Rebel newspaper, Vicksburg, Miss.; July 2, 1863. *Vicksburg Daily Citizen*, June 20, 1863. Schoolcraft's N. A. Indians. Report on Forestry, U. S. Document.

www.ingramcontent.com/pod-product-compliance
Lightning Source LLC
Chambersburg PA
CBHW021936190326
41519CB00009B/1031